MATRIX THEORY
AND ITS
APPLICATIONS

SELECTED TOPICS

PURE AND APPLIED MATHEMATICS

A Program of Monographs, Textbooks, and Lecture Notes

MONOGRAPHS AND TEXTBOOKS IN PURE AND APPLIED MATHEMATICS

1. K. YANO. Integral Formulas in Riemannian Geometry (1970)
2. S. KOBAYASHI. Hyperbolic Manifolds and Holomorphic Mappings (1970)
3. V. S. VLADIMIROV. Equations of Mathematical Physics (A. Jeffrey, editor; A. Littlewood, translator) (1970)
4. B. N. PSHENICHNYI. Necessary Conditions for an Extremum (L. Neustadt, translation editor; K. Makowski, translator) (1971)
5. L. NARICI, E. BECKENSTEIN, and G. BACHMAN. Functional Analysis and Valuation Theory (1971)
6. D. S. PASSMAN. Infinite Group Rings (1971)
7. L. DORNHOFF. Group Representation Theory (in two parts). Part A: Ordinary Representation Theory. Part B: Modular Representation Theory (1971, 1972)
8. W. BOOTHBY and G. L. WEISS (eds.). Symmetric Spaces: Short Courses Presented at Washington University (1972)
9. Y. MATSUSHIMA. Differentiable Manifolds (E. T. Kobayashi, translator) (1972)
10. L. E. WARD, JR. Topology: An Outline for a First Course (1972)
11. A. BABAKHANIAN. Cohomological Methods in Group Theory (1972)
12. R. GILMER. Multiplicative Ideal Theory (1972)
13. J. YEH. Stochastic Processes and the Wiener Integral (1973)
14. J. BARROS-NETO. Introduction to the Theory of Distributions (1973)
15. R. LARSEN. Functional Analysis: An Introduction (1973)
16. K. YANO and S. ISHIHARA. Tangent and Cotangent Bundles: Differential Geometry (1973)
17. C. PROCESI. Rings with Polynomial Identities (1973)
18. R. HERMANN. Geometry, Physics, and Systems (1973)
19. N. R. WALLACH. Harmonic Analysis on Homogeneous Spaces (1973)
20. J. DIEUDONNÉ. Introduction to the Theory of Formal Groups (1973)
21. I. VAISMAN. Cohomology and Differential Forms (1973)
22. B.-Y. CHEN. Geometry of Submanifolds (1973)
23. M. MARCUS. Finite Dimensional Multilinear Algebra (in two parts) (1973, 1975)
24. R. LARSEN. Banach Algebras: An Introduction (1973)
25. R. O. KUJALA and A. L. VITTER (eds). Value Distribution Theory: Part A; Part B. Deficit and Bezout Estimates by Wilhelm Stoll (1973)
26. K. B. STOLARSKY. Algebraic Numbers and Diophantine Approximation (1974)
27. A. R. MAGID. The Separable Galois Theory of Commutative Rings (1974)
28. B. R. McDONALD. Finite Rings with Identity (1974)
29. I. SATAKE. Linear Algebra (S. Koh, T. Akiba, and S. Ihara, translators) (1975)
30. J. S. GOLAN. Localization of Noncommutative Rings (1975)
31. G. KLAMBAUER. Mathematical Analysis (1975)
32. M. K. AGOSTON. Algebraic Topology: A First Course (1976)
33. K. R. GOODEARL. Ring Theory: Nonsingular Rings and Modules (1976)
34. L. E. MANSFIELD. Linear Algebra with Geometric Applications (1976)
35. N.J. PULLMAN. Matrix Theory and its Applications: Selected Topics (1976)

MATRIX THEORY AND ITS APPLICATIONS

SELECTED TOPICS

N. J. Pullman
Department of Mathematics
Queen's University at Kingston
Ontario, Canada

MARCEL DEKKER, INC.   New York and Basel

QA188
P84

MARCEL DEKKER, INC.
270 Madison Avenue, New York, New York 10016

LIBRARY OF CONGRESS CATALOG CARD NUMBER: 75-40845

ISBN: 0-8247-6420-X

Current printing (last digit):
10 9 8 7 6 5 4 3 2 1

PRINTED IN THE UNITED STATES OF AMERICA

PREFACE

This book is based on notes developed over a period of five
years for a half-course given for third and fourth year
undergraduates and beginning graduate students in several
disciplines. It assumes an introductory course in linear
algebra and a second course in analysis.

The main objective is to help the student become sufficiently fluent in matrix theory to enable him to use it
effectively in his own area of interest.

After an introductory chapter reviewing certain areas
of matrix algebra and presenting some basic ideas of matrix
analysis, three topics are presented: nonnegative matrices,
differential equations, and location of eigenvalues. These
topics were chosen primarily because we felt that all of
our students, despite their varied interests (economics,
electrical engineering, statistics, etc.), could obtain
enough of an impression of the applications of these topics
to appreciate the development of the theory.

Many of the exercises, examples, and review topics that
appear in the book are there in response to classroom experience.

It is hoped that the book will give the student enough
exposure to the development of matrix-theoretic ideas so

that when faced with problems in his own discipline he may
know when and how to use the matrix theory he has learned,
and perhaps be able to improvise new methods when no ready-made
ones are available.

   Much of the material presented here was gathered from
two principal sources, both of which are close to being
encyclopedias of matrix theory: "Applications of the Theory
of Matrices," vols. I and II, by F. R. Gantmacher and "A Survey of Matrix Theory and Matrix Inequalities" by M. Marcus
and H. Minc. The details of these and other sources are
given at the end of the book.

   I am indebted to Professor Marvin Marcus for his encour-
agement and constructuve criticism.

   Professor Robert M. Erdahl deserves thanks for his con-
tributions and suggestions.

   I am grateful to Mmes. Diane Quesnelle, Karen Lewis,
and Marge Lambert for their intrepid typing, to Mr. James
Carswell for his vigilant proofreading and helpful comments,
to Mr. Shahid Jamil for his work on the index, and to Mrs.
Ruth Cross Park who suggested that this be written.

                                            N. J. Pullman

CONTENTS

MATRIX THEORY
AND ITS
APPLICATIONS

SELECTED TOPICS

CHAPTER 1

ALGEBRAIC AND ANALYTIC PRELIMINARIES

1. NOTATION

We shall reserve all the capital Latin Letters (A, B, etc.)
for matrices, the letters u, v, w, x, y, z for column
vectors and reserve the letters i, j, k, ℓ, m, n for
integer indices (except when $i = \sqrt{-1}$). The letters f, g, h
are reserved for functions. The entries in a matrix A (or
a vector x) are denoted by the same letter with the
appropriate subscripts (a_{ij} or x_j). Other scalars are
denoted by lowercase Greek letters (α, β, etc.). Unless we
specify otherwise, when we say *matrix* we mean complex,
k × k matrix, when we say *scalar* (or *number*) we mean complex
number, and usually when we say *vector* we mean a complex
k × 1 matrix (a complex *column* vector). Other notation
will be explained as we go along. These are collected in
the Notation at the end of the book.

2. BACKGROUND

You are expected to recall, or at any rate to review,
certain basic ideas of matrix theory: initially you should
make sure you know the meaning of *eigenvalue*, *eigenvector*
and *characteristic polynomial*. This involves remembering

1

what the *determinant* of a matrix is, reviewing its
properties, and, in particular, recalling its connection
with the question of whether or not a given matrix is
singular. You should also be familiar with the ideas of
vector space and *subspace, linear dependence* and
independence, linear combinations of vectors, the space
spanned by a set of vectors, the *basis* of a vector space
and its *dimension* and *linear mapping* and *linear operator* as
given in most standard first courses in linear algebra.
When the occasion arises, references to a standard textbook
or reference book will be given. The details of the
reference (title, publisher, etc.) will be found in the
Reference section at the end of the book. Suggestions for
collateral reading are also given there.

    You are also expected to have had some experience
with elementary real analysis (and to a somewhat lesser
extent with complex analysis) at least to being able to
deal with infinite series and sequences.

3.   SIMILARITY AND SIMILARITY INVARIANTS

A matrix A is said to be *similar* to a matrix B (written:
$A \sim B$) if and only if there is a matrix P such that $A = PBP^{-1}$
Geometrically speaking, $A \sim B$ means that A and B both
represent the same linear operator. You should be able to
see why $\sim$ is an equivalence relation (see Exercise 1).

Similarity Invariants

A function f defined on matrices is said to be a
similarity invariant if f(A) = f(B), whenever A ~ B. Such
functions cannot distinguish between similar matrices. For
example, the determinant, trace, and rank are similarity
invariants.

The characteristic polynomial is a similarity
invariant (it too is a function with a matrix argument, but
it takes on polynomial values). The minimal polynomial is
another polynomial-valued similarity invariant.

A set of matrices is said to be *similarity invariant*
(or *invariant under similarity*) if all matrices similar to
a matrix in the set are also in the set. For instance,
given α, β, and γ, the set of all solutions X to
$\alpha X^2 + \beta X + \gamma I = 0$ is invariant under similarity (see
Exercise 5). A property is said to be *invariant under*
similarity (or *similarity invariant*) if the set of matrices
having the property is similarity invariant; for example,
invertibility is a similarity invariant.

Exercises

1. Prove that similarity is an equivalence relation, i.e.,
    that for all A, B, and C: A ~ A; if A ~ B, then B ~ A;
    if A ~ B and B ~ C, then A ~ C.
2. a. If I is the identity matrix, show that λI is the
        only matrix similar to λI.

b. Show that $A = \lambda I$ if A is the only matrix similar to
A. Try to do this without using Sec. 6. (*Hint:*
If A is the only matrix similar to A, then $PA = AP$
for *all* invertible matrices P. Thus, for example,
if $A = \begin{bmatrix} \lambda & \beta \\ \gamma & \delta \end{bmatrix}$, then $\begin{bmatrix} 0 & 1 \\ 1 & 0 \end{bmatrix} A = A \begin{bmatrix} 0 & 1 \\ 1 & 0 \end{bmatrix}$,

$\begin{bmatrix} \gamma & \delta \\ \lambda & \beta \end{bmatrix} = \begin{bmatrix} \beta & \lambda \\ \delta & \gamma \end{bmatrix}$, and $\begin{bmatrix} 0 & 1 \\ -1 & 0 \end{bmatrix} A = A \begin{bmatrix} 0 & 1 \\ -1 & 0 \end{bmatrix}$; so

$\begin{bmatrix} \gamma & \delta \\ -\lambda & -\beta \end{bmatrix} = \begin{bmatrix} -\beta & \lambda \\ -\delta & \gamma \end{bmatrix}$. Therefore $\beta = \gamma = -\beta$ and

$\lambda = \delta$; hence $\beta = \gamma = 0$ and $A = \lambda I$. Now extend this
method so it will work for matrices of arbitrary
order.)

3. a. If g is a *matrix polynomial* (that is,
$g(T) = \alpha_0 I + \alpha_1 T + \alpha_2 T^2 + \cdots + \alpha_n T^n$ for all square
matrices T) and $A = PBP^{-1}$, show that $g(A) = Pg(B)P^{-1}$.
[*Hint:* $\alpha_2 A^2 = \alpha_2 (PBP^{-1})^2 = \alpha_2 PB(P^{-1}P)BP^{-1} = \alpha_2 PB^2 P^{-1}$.]

b. If D is a *diagonal* matrix (that is, $d_{ij} = 0$ if
$i \neq j$), show that g(D) is a diagonal matrix whose
ii^{th} entry is $g(d_{ii})$.

4. Prove that the determinant and trace of a matrix are
similarity invariants.

5. a. Given α, β, and γ, suppose that $\Lambda = \{X : \alpha X^2 + \beta X + \gamma I = 0\}$. Show that Λ is similarity invariant.

b. Are there α, β, and γ such that $\Lambda = \phi$?

6. A matrix A is said to be *nilpotent* iff $A^\ell = 0$ for some ℓ.

a. Show that nilpotency is a similarity invariant.

   b. The first $\ell > 0$ such that $A^\ell = 0$ is called the
       index of nilpotency of A. Show that this index
       is also a similarity invariant.

7. A is *idempotent* iff $A^2 = A$. Show that idempotence is
    a similarity invariant.

8. Show that the minimal polynomial is a similarity
    invariant. (Suggestion: The minimal polynomial of A
    divides every polynomial which has the matrix A for a
    root.) (See Sec. 17 for a review of minimal polynomial.)

4. HOW IS SIMILARITY USED IN SOLVING PROBLEMS?

The Similarity Method

     There is a strategy for solving problems that is used
frequently enough to deserve some special attention. We'll
call this technique "the similarity method." Here is an
outline of the method:

Step 1. Choose a matrix B similar to A for which the
         problem is easier to solve.

Step 2. Solve the problem using the matrix B instead of A
         (the B-problem).

Step 0. Interpret the solution to the B-problem in terms
         of the matrix A.

     Example: Given $A = \begin{bmatrix} 1 & 4 \\ 3 & 2 \end{bmatrix}$, find (each entry in) A^{1010}.

Step 1. (Choose easier B $\sim$ A)

         Choose $B = \begin{bmatrix} 5 & 0 \\ 0 & -2 \end{bmatrix}$, where $P = \begin{bmatrix} 1 & -4 \\ 1 & 3 \end{bmatrix}$; then $A = PBP^{-1}$.

Step 2.   (Solve B-problem)

$$B^{1010} = \begin{bmatrix} 5^{1010} & 0 \\ 0 & (-2)^{1010} \end{bmatrix}$$

Step 3.   (Interpret B-solution)

$$A^{1010} = (PBP^{-1})^{1010}$$
$$= PB^{1010}P^{-1} \quad \text{(see Exercise 3)}$$

Thus $A^{1010} = \dfrac{1}{7} \begin{bmatrix} 3(5^{1010})+2^{1012} & 4(5^{1010})-2^{1012} \\ 3(5^{1010})-3(2^{1010}) & 4(5^{1010})+3(2^{1010}) \end{bmatrix}$

As you have probably observed already, the first obstacle to using the similarity method is that it may not be clear which B to choose. Indeed, there may be no B similar to A which makes the problem easier. Nevertheless, there are many problems which can be attacked successfully by this method. This is particularly true when A is "diagonable."

5.   DIAGONABLE MATRICES

When a matrix is *diagonable* (i.e., when it is similar to a diagonal matrix), then the similarity method frequently works, since choosing a diagonal matrix B in Step 1 will probably make Step 2 easy because diagonal matrices are so simple. Step 3 may sometimes turn out to be difficult if not impossible (see Sec. 7 for an illustration), but the previous example and the following example are instances where Step 3 can be carried out easily.

Example: In certain problems in economics the state of
a system is described by a matrix $S_n = I + A + A^2 + \cdots + A^n$
at time n where A is a given matrix. (We'll go into the
background of this problem in more detail later on.) We
are interested in the system "in the long run," i.e., when
n is large. Suppose $A = \begin{bmatrix} 0.1 & 0.7 \\ 0.3 & 0.6 \end{bmatrix}$. Fortunately, as in
the first example, A is diagonable because it is a 2 × 2
matrix having two distinct eigenvalues.[†] Calling these
values λ and μ we know that $A = P\begin{bmatrix} \lambda & 0 \\ 0 & \mu \end{bmatrix}P^{-1}$ for some matrix
P[††]. (You should be able to calculate λ, μ, and P. [see
e.g., Lipschutz (1968), pp. 198-200].)

Step 1. (Choose B ~ A)

So we choose $B = \begin{bmatrix} \lambda & 0 \\ 0 & \mu \end{bmatrix}$.

Step 2. (Solve B-problem)

Let $T_n = I + B + B^2 + \cdots + B^n$, then

$$T_n = \begin{bmatrix} 1 + \lambda + \lambda^2 + \cdots + \lambda^n & 0 \\ 0 & 1 + \mu + \mu^2 + \cdots + \mu^n \end{bmatrix}$$

$$= \begin{bmatrix} \dfrac{1 + \lambda^{n+1}}{1 - \lambda} & 0 \\ 0 & \dfrac{1 - \mu^{n+1}}{1 - \mu} \end{bmatrix} \quad \text{for all } n \geq 0$$

Since $|\lambda| < 1$ and $|\mu| < 1$, it follows that T_n

[†] You may recall that a k × k matrix is diagonable if it has
k distinct eigenvalues. The converse isn't true.

[††] Take P to be the matrix whose j^{th} column is an eigenvector
for the jth eigenvalue.

approximates $\begin{bmatrix} \frac{1}{1-\lambda} & 0 \\ 0 & \frac{1}{1-\mu} \end{bmatrix}$ when n is large. Call

this latter matrix T.

Step 3.   (Interpret B-solution)

Since $S_n = PT_nP^{-1}$, it follows that S_n is

approximately PTP^{-1} for large enough n (how good

an approximation it is depends on how large n is)

and so even without computing P we can predict

that in the long run the system will "reach a

steady state," i.e., S_n isn't appreciably different

from S_m when n and m are sufficiently large. If

we want a quantitative statement, we calculate P

and can then say how large n has to be for S_n to

approximate PTP^{-1} (the steady-state matrix) within,

say, two-decimal place accuracy or whatever accuracy

is required.

Exercises

9.   a.   Find λ, μ, and P of the previous example.

     b.   How large must n be for each entry in S_n to

          approximate the corresponding entry in PTP^{-1} to two

          decimal places (S_n, P, T as in previous example)?

10.   Prove that every diagonable matrix is similar to its

      transpose. (Later on we'll be able to drop the

      diagonability requirement.) (*Hint:* Use the similarity

      method with B the diagonal matrix similar to A.)

11. If A is any diagonable matrix and $g(\tau) = \alpha_0 + \alpha_1\tau + \cdots +$
$\alpha_{k-1}\tau^{k-1} + \tau^k$ is the characteristic polynomial of A, show
that $g(A) = 0$ (i.e., $\alpha_0 I + \alpha_1 A + \cdots + \alpha_{k-1}A^{k-1} + A^k = 0$).
(This is the Cayley-Hamilton theorem. It's true even if
A isn't diagonable; we'll see why later on.) [*Hint:* If
B is a diagonal matrix similar to A, use Exercise 3(b)
and the fact that the diagonal entries of B are eigenvalues of A to show that $g(B) = 0$.

12. If $x_0 = 1$, $x_1 = 2$, and $x_n = x_{n-1} + 2x_{n-2}$ for all
$n \geq 2$, find x_n as a function of n. (Suggestion: Let
$A = \begin{bmatrix} 0 & 1 \\ 2 & 1 \end{bmatrix}$, then $A\begin{bmatrix} x_0 \\ x_1 \end{bmatrix} = \begin{bmatrix} x_1 \\ x_1 + 2x_0 \end{bmatrix} = \begin{bmatrix} x_1 \\ x_2 \end{bmatrix}$,

$A\begin{bmatrix} x_1 \\ x_2 \end{bmatrix} = \begin{bmatrix} x_2 \\ x_2 + 2x_1 \end{bmatrix} = \begin{bmatrix} x_2 \\ x_3 \end{bmatrix}$; so $A^2\begin{bmatrix} x_0 \\ x_1 \end{bmatrix} = \begin{bmatrix} x_2 \\ x_3 \end{bmatrix}$, and in

general, $A^n\begin{bmatrix} x_0 \\ x_1 \end{bmatrix} = \begin{bmatrix} x_n \\ x_{n+1} \end{bmatrix}$ for all n.) (See Exercise

14 for a shortcut to avoid computing P.)

13. If $B_n = \begin{bmatrix} 1 & 1 & 0 & 0 & 0 & 0 & \cdots & 0 & 0 & 0 & 0 & 0 \\ 1 & 1 & 1 & 0 & 0 & 0 & \cdots & 0 & 0 & 0 & 0 & 0 \\ 0 & 1 & 1 & 1 & 0 & 0 & \cdots & 0 & 0 & 0 & 0 & 0 \\ 0 & 0 & 1 & 1 & 1 & 0 & \cdots & 0 & 0 & 0 & 0 & 0 \\ \cdots & & & & & & & & & & & \\ 0 & 0 & 0 & 0 & 0 & 0 & \cdots & 0 & 1 & 1 & 1 & 0 \\ 0 & 0 & 0 & 0 & 0 & 0 & \cdots & 0 & 0 & 1 & 1 & 1 \\ 0 & 0 & 0 & 0 & 0 & 0 & \cdots & 0 & 0 & 0 & 1 & 1 \end{bmatrix}$, find $\det(B_n)$.

[Suggestion: Expand $\det(B_n)$ by minors of the first
column to obtain a relation between $\det(B_n)$, $\det(B_{n-1})$,
and $\det(B_{n-2})$, then apply the technique of Exercise
12.]

14. If A is diagonable and $A^n[x_0,\ldots, x_{k-1}]^{tr} =$
     $[x_n,\ldots, x_{n+k-1}]^{tr}$, show that there are scalars
     β_1, $\beta_2,\ldots, \beta_\ell$ (independent of n) such that
     $x_n = \sum_{j=1}^{\ell} \beta_j \lambda_j^n$, where λ_1, $\lambda_2,\ldots, \lambda_\ell$ are the distinct

     eigenvalues of Λ. (This observation enables you to
     solve problems such as 12 and 13 without finding a
     matrix P which diagonalizes A as you need only
     determine the β_i by using the first ℓ values of x_n.)
15. If A is diagonable, find necessary and sufficient
     conditions on the eigenvalues of A for A to be
     idempotent (see Exercise 7).

6. JORDAN MATRICES

Unfortunately, not all matrices are diagonable. [For
example, the matrix $\begin{bmatrix} 1 & 1 \\ 0 & 1 \end{bmatrix}$ isn't diagonable, for if it
were , then $\begin{bmatrix} 1 & 1 \\ 0 & 1 \end{bmatrix} \sim \begin{bmatrix} \lambda & 0 \\ 0 & \mu \end{bmatrix}$ for some λ, μ and λ, μ must
both be 1. (Why?) Therefore, $\begin{bmatrix} 1 & 1 \\ 0 & 1 \end{bmatrix} \sim \begin{bmatrix} 1 & 0 \\ 0 & 1 \end{bmatrix}$; hence
$\begin{bmatrix} 1 & 1 \\ 0 & 1 \end{bmatrix} = \begin{bmatrix} 1 & 0 \\ 0 & 1 \end{bmatrix}$ (Why?) which would imply that 1 = 0.]

   However, there are many questions about arbitrary
matrices A which can be treated by the similarity method if
in Step 1 we choose a Jordan matrix B which is similar to
A. These matrices are almost as easy to deal with as
diagonal matrices.

Some Preliminary Definitions

A *Jordan block* $J_n(\lambda)$ is an $n \times n$ matrix:

1. Each of whose diagonal entries is λ

2. Each of whose superdiagonal entries is 1

3. Each of whose other entries is zero

Thus
$$J_4(2) = \begin{bmatrix} 2 & 1 & 0 & 0 \\ 0 & 2 & 1 & 0 \\ 0 & 0 & 2 & 1 \\ 0 & 0 & 0 & 2 \end{bmatrix}$$

and $J_1(2) = 2$ (or $[2]$ if you are a purist).

The next idea we need is the direct sum of matrices: If A is an $n \times n$ matrix and B is an $m \times m$ matrix then the *direct sum* $A \oplus B$ is the $(n + m) \times (n + m)$ matrix C for which

$$c_{ij} = a_{ij} \quad \text{for all} \quad 1 \le i, j \le n$$
$$c_{ij} = b_{ij} \quad \text{for all} \quad n + 1 \le 1, j \le n + m$$
$$c_{ij} = 0 \quad \text{for all other } i, j$$

Thus
$$\begin{bmatrix} 1 & 2 \\ 3 & 4 \end{bmatrix} \oplus \begin{bmatrix} 1 & 0 & 2 \\ 1 & -1 & 1 \\ 0 & 1 & 0 \end{bmatrix} = \begin{bmatrix} 1 & 2 & 0 & 0 & 0 \\ 3 & 4 & 0 & 0 & 0 \\ 0 & 0 & 1 & 0 & 2 \\ 0 & 0 & 1 & -1 & 1 \\ 0 & 0 & 0 & 1 & 0 \end{bmatrix} \text{ and}$$

$$2 \oplus \begin{bmatrix} 1 & 3 \\ 0 & 4 \end{bmatrix} = \begin{bmatrix} 2 & 0 & 0 \\ 0 & 1 & 3 \\ 0 & 0 & 4 \end{bmatrix}$$

We extend this definition to cover three summands by defining $A \oplus B \oplus C = (A \oplus B) \oplus C$. In general, we extend it inductively to cover n summands by defining

$$A_1 \oplus A_2 \oplus \cdots \oplus A_{n+1} = (A_1 \oplus A_2 \oplus \cdots \oplus A_n) \oplus A_{n+1}$$

To save space we'll write $\displaystyle\overset{n}{\underset{i=1}{\oplus}} A_i$ for $A_1 \oplus A_2 \oplus \cdots \oplus A_n$.

Thus $\overset{3}{\underset{i=1}{\oplus}} [i] = \begin{bmatrix} 1 & 0 & 0 \\ 0 & 2 & 0 \\ 0 & 0 & 3 \end{bmatrix}$, $\overset{n}{\underset{i=1}{\oplus}} [1]$ is the n × n identity

matrix, and

$$\overset{3}{\underset{i=1}{\oplus}} \begin{bmatrix} \lambda_i & 1 \\ 0 & \lambda_i \end{bmatrix} = \begin{bmatrix} \lambda_1 & 1 & 0 & 0 & 0 & 0 \\ 0 & \lambda_1 & 0 & 0 & 0 & 0 \\ 0 & 0 & \lambda_2 & 1 & 0 & 0 \\ 0 & 0 & 0 & \lambda_2 & 0 & 0 \\ 0 & 0 & 0 & 0 & \lambda_3 & 1 \\ 0 & 0 & 0 & 0 & 0 & \lambda_3 \end{bmatrix}$$

Exercises

16. Suppose $g(T) = \overset{n}{\underset{i=0}{\Sigma}} \alpha_i T^i$, for all k × k matrices T and

for all $k \geq 1$. Show that $g \overset{m}{\underset{j=1}{\oplus}} T_j = \overset{m}{\underset{j=1}{\oplus}} g(T_j)$.

17. Show that $\overset{m}{\underset{j=1}{\oplus}} T_j$ is nilpotent iff each T_j is nilpotent.

(See Exercise 6.)

18. Show that $\overset{m}{\underset{j=1}{\oplus}} T_j$ is idempotent iff each T_j is

idempotent. (See Exercise 7.)

We say that J is a *Jordan matrix* iff J is a direct

sum of Jordan blocks. Thus

$$\begin{bmatrix} 2 & 1 \\ 0 & 2 \end{bmatrix} \oplus 2 \oplus \begin{bmatrix} 4 & 1 & 0 \\ 0 & 4 & 1 \\ 0 & 0 & 4 \end{bmatrix} = \begin{bmatrix} 2 & 1 & 0 & 0 & 0 & 0 \\ 0 & 2 & 0 & 0 & 0 & 0 \\ 0 & 0 & 2 & 0 & 0 & 0 \\ 0 & 0 & 0 & 4 & 1 & 0 \\ 0 & 0 & 0 & 0 & 4 & 1 \\ 0 & 0 & 0 & 0 & 0 & 4 \end{bmatrix}$$

$$= J_2(2) \oplus J_1(2) \oplus J_3(4)$$

is a Jordan matrix. Another way to write this one is

$\bigoplus_{i=1}^{3} J_{n_i}(\lambda_i)$, where $n_1 = 2$, $n_2 = 1$, $n_3 = 3$ and $\lambda_1 = 2$,

$\lambda_2 = 2$, $\lambda_3 = 4$. In general, J is a Jordan matrix iff

$J = \bigoplus_{i=1}^{m} J_{n_i}(\lambda_i)$. If J is $k \times k$ then $\sum_{i=1}^{m} n_i = k$. It's also

easy to see that the characteristic polynomial of J is

$(\tau - \lambda_1)^{n_1} (\tau - \lambda_2)^{n_2} \cdots (\tau - \lambda_m)^{n_m}$. [In our example, the

polynomial is $(\tau - 2)^2 (\tau - 2)^1 (\tau - 4)^3$.] Thus the

multiplicity of the eigenvalue   (i.e., the maximum n such

that $(\tau - \lambda)^n$ is a factor of the characteristic polynomial)

is the sum of those n_i for which $\lambda_i = \lambda$. (In our example,

the multiplicity of the eigenvalue 2 is three and

$n_1 + n_2 = 2 + 1 = 3$.)

7. USING THE SIMILARITY METHOD ON ARBITRARY MATRICES: THE
    CAYLEY-HAMILTON THEOREM, EVALUATION OF MATRIX POLYNOMIALS

We said at the outset of Sec. 6 that Jordan matrices are

almost as easy to use as diagonal matrices and that many

problems concerning matrices can be attacked by the similarity

method if we choose a Jordan matrix similar to A at Step 1.

Having seen that not all matrices are diagonable:

1. How can we tell whether there is a Jordan matrix

    similar to A to choose? There are tests for

    diagonability (e.g., if all eigenvalues of A have

    multiplicity one, then A is diagonable).

2. What are the tests for "Jordanability"?

The answer to question 1 was given by Camille Jordan; it

neatly disposes of question (2) also.

Jordan's Theorem.[†] Every matrix is similar to some Jordan matrix.

So we see that we can always choose a Jordan matrix (it will be diagonal if A were diagonable) at Step 1, and often the "B-problem" can be solved for these matrices. But, as we mentioned in Sec. 4, we may still encounter an insurmountable obstacle at Step 3. For example: Is it true that $\max\limits_{1 \leq i \leq k} \left(\sum\limits_{j=1}^{k} |a_{ij}| \right) \leq |\lambda|$ for all eigenvalues λ when A is an arbitrary k × k matrix?

If we choose B to be a Jordan matrix, then the question is answered easily for the matrix B (the answer is yes). But you will find that it is not so easy to perform the third step of the similarity method in this case. This is mainly because the function f(A) = $\max\limits_{1 \leq i < k} \sum\limits_{j=1}^{k} |a_{ij}|$ is not similarity invariant. Although the answer to the original question turns out to be "yes," we need to use a technique other than the similarity method to prove it because of this difficulty with Step 3.

To summarize: Jordan's Theorem enables us to get as far as Step 1 of the similarity method using a Jordan matrix for B no matter what matrix A we begin with. Usually Step 2 is easy (because Jordan matrices have so

---

[†] For a proof see e.g., Hoffman and Kunze (1961, Secs. 7.2 and 7.3).

many zeroes), but Step 3 may be difficult if not impossible
to perform because of the nature of the problem.

As an application of the similarity method we shall
prove the extremely useful Cayley-Hamilton theorem.

Cayley-Hamilton Theorem. If $g(\tau)$ is the characteristic
polynomial of an arbitrary matrix A, then $g(A) = 0$. That
is, if $g(\tau) = \alpha_0 + \alpha_1\tau + \cdots + \alpha_{k-1}\tau^{k-1} + \tau^k$ is the
characteristic polynomial of A, then $\alpha_0 I + \alpha_1 A + \alpha_2 A^2 + \cdots +$
$\alpha_{k-1}A^{k-1} + A^k = 0$.

Proof: $A = PJP^{-1}$, where J is a Jordan matrix
$\overset{m}{\underset{i=1}{\oplus}} J_{n_i}(\lambda_i)$. Therefore by Exercises 3(a) and 16,

$$g(A) = \left[P \overset{m}{\underset{i=1}{\oplus}} g\, J_{n_i}(\lambda_i) \right] P^{-1}$$

So it is sufficient to prove that $g\, J_{n_i}(\lambda_i) = 0$ for each

$1 \le i \le m$. Now $\overset{m}{\underset{i=1}{\prod}} (\tau - \lambda_i)^{n_i}$, as we observed in Sec. 6, is

the characteristic polynomial for J, and hence by Exercise 4,

$$g(\tau) = \overset{m}{\underset{i=1}{\prod}} (\tau - \lambda_i)^{n_i}. \quad \text{If we let } g_j(\tau) = \overset{m}{\underset{\substack{i=1 \\ i \ne j}}{\prod}} (\tau - \lambda_i)^{n_i},$$

then $g(\tau) = g_j(\tau)(\tau - \lambda_j)^{n_j}$, and hence by Exercise 20 we
have $g(T) = g_j(T)(T - \lambda_j I_{n_j})^{n_j}$ for every square matrix

T; in particular, for $T = J_{n_j}(\lambda_j)$, but then $(T - \lambda_j I_{n_j})^{n_j} =$

$(J_{n_j}(0))^{n_j} = 0$ [Exercise 22(a)], and hence $g(J_{n_j}(\lambda_j)) = 0$
for each $1 \le j \le m$. []

Exercise 20 we have $g(T) = g_j(T)(T - \lambda_j I_{n_j})^{n_j}$ for every

square matrix T; in particular, for $T = J_{n_j}(\lambda_j)$, but then

$(T - \lambda_j I_{n_j})^{n_j} = (J_{n_j}(0))^{n_j} = 0$ [Exercise 22(a)], and hence

$g(J_{n_j}(\lambda_j)) = 0$ for each $1 \leq j \leq m$. []

Exercises

19. Prove that every matrix is similar to its transpose.

(*Hint:* If $T_3 = \begin{bmatrix} 0 & 0 & 1 \\ 0 & 1 & 0 \\ 1 & 0 & 0 \end{bmatrix}$, then $T_3^{-1} = T_3$ and $T_3 J_3(\lambda) T_3 =$

$= [J_3(\lambda)]^{tr}$. For each n, construct a T_n such that

$T_n^{-1} J_n(\lambda) T_n = [J_n(\lambda)]^{tr}$, then use the similarity method.)

20. Suppose f, g, and h are polynomials (with scalar

coefficients).

a.   If $f(\tau) = g(\tau) + h(\tau)$ for all scalars τ, show that

$f(T) = g(T) + h(T)$ for all square matrices T.

b.   If $f(\tau) = g(\tau)h(\tau)$ for all scalars τ, show that

$f(T) = g(T)h(T)$ for all square matrices T.

Solution a.:   Suppose $f(\tau) = \sum_{i=0}^{n} \alpha_i \tau^i$, $g(\tau) = \sum_{i=0}^{n} \beta_i \tau^i$,

and $h(\tau) = \sum_{i=0}^{n} \gamma_i \tau^i$. We have $\alpha_i = \beta_i + \gamma_i$ $(0 \leq i \leq n)$,

because $f(\tau) = g(\tau) + h(\tau)$ for all τ but

$f(T) = \sum_{i=0}^{n} \alpha_i T^i$ by definition (see Exercise 3)

$= \sum_{i=0}^{n} (\beta_i + \gamma_i) T^i$

$= \left(\sum_{i=0}^{n} \beta_i T^i \right) + \left(\sum_{i=0}^{n} \gamma_i T^i \right)$

$= g(T) + h(T)$ (Why?)

Solution b.:   Let $g_i(\tau) = \alpha_i \tau^i$, $h_j(\tau) = \beta_j \tau^j$, and $f_{ij}(\tau) = g_i(\tau)h_j(\tau)$.   We have

$$f(\tau) = g(\tau)h(\tau) = \sum_{i=0}^{n} \sum_{j=0}^{n} g_i(\tau)h_j(\tau) = \sum_{i=0}^{n} \sum_{j=0}^{n} f_{ij}(\tau)$$

and hence

$$f(T) = \sum_{i=0}^{n} \sum_{j=0}^{n} f_{ij}(T) \quad \text{by (a)}$$

But $f_{ij}(\tau) = \alpha_i \beta_j \tau^{i+j}$, thus

$$f_{ij}(T) = \alpha_i \beta_j T^{i+j} = \alpha_i T^i \beta_j T^j = g_i(T)h_j(T)$$

and hence

$$f(T) = \sum_{i=0}^{n} \sum_{j=0}^{n} g_i(T)h_j(T)$$

$$= \sum_{i=0}^{n} g_i(T) \sum_{j=0}^{n} h_j(T)$$

$$= g(T)h(T)$$

c.   You may recall the <u>Scalar Binomial Theorem</u>:

$$(\lambda + \tau)^n = \sum_{j=0}^{n} \binom{n}{j} \lambda^j \tau^{n-j}, \quad \text{for scalars } \lambda \text{ and } \tau \quad [\binom{n}{j}$$

denotes $n!/j!(n-j)!$ for each $0 \le j \le n$].

Prove the *Matrix Binomial Theorem*:

$$(S + T)^n = \sum_{j=0}^{n} \binom{n}{j} S^j T^{n-j}$$

for all square matrices S, T such that ST = TS.

21. If A is invertible and k × k prove that there exist
     scalars γ_j such that $A^{-1} = \sum_{j=0}^{k-1} \gamma_j A^j$. (*Hint*: If A^2 +

     $\alpha A + \beta I = 0$ and $\beta \neq 0$ then $A(A + \alpha I)\beta^{-1} = -I$ and hence
     $A^{-1} = -\beta^{-1}A - (\alpha\beta^{-1})I$.)

22. A characterization of nilpotence:

     a. Show that $(J_n(0))^n = 0$.

     b. Find a condition on the eigenvalues of A which is
         equivalent to the nilpotency of A. (Suggestion:
         use the similarity method and Exercises 6 and 17.

     c. Prove that the condition stated in (b) is equivalent
         to the nilpotency of A.

23. A characterization of idempotence:

     a. Under what conditions on n and λ is $J_n(\lambda)$ idempotent?

     b. State a set of conditions on an arbitrary matrix A
         equivalent to idempotence (see Exercises 7, 15, and
         18).

     c. Prove the statement made in (b).

24. Characterize those matrices X such that $X^3 = I$.

25. If $x_0 = 1$, $x_1 = 1$, and $x_n = 2x_{n-1} - 4x_{n-2}$ for all $n \geq 2$,
     write x_n as a function of n. (See Exercise 12.)

26. If $A^n[x_0,x_1,\ldots,x_{k-1}]^{tr} = [x_n,x_{n+1},\ldots,x_{n+k-1}]^{tr}$, show
     that there exist scalars β_{ij} (independent of n) such
     that

$$x_n = \sum_{i=1}^{m} \sum_{j=0}^{n_i} \beta_{ij} \binom{n}{j} \lambda_i^{n-j} \quad \text{(see Exercise 14)}$$

27. Express the minimal polynomial of A in terms of the λ_i
     and n_i where $\bigoplus_{i=1}^{m} J_{n_i}(\lambda_i)$ is similar to A.

Two Applications of the Cayley-Hamilton Theorem

The matrix equation $AX = XB$. The question of whether
the equation $AX = XB$ has a nonzero solution X turns out to
be important in many different contexts. We shall need the
following lemma twice in the work to come.

Lemma 7.1. Suppose A is $k \times k$ and B is $\ell \times \ell$. If A
has no eigenvalue in common with B, then the equation $AX =$
XB has only one solution, $X = 0$.

Proof: Suppose $AX = XB$. If $c(\tau)$ is the characteristic
polynomial of B, then $c(A)X = Xc(B)$ (see Exercise 28), and
hence $c(A)X = 0$ by the Cayley-Hamilton theorem. But $c(A)$ is
invertible because A has no eigenvalues in common with B
(see Exercise 29), therefore $X = 0$. []

Exercises

(*Note:* In the following, A is $k \times k$ and B is $\ell \times \ell$.)

28. If p is a polynomial and $AX = XB$, show that $p(A)X =$
     $Xp(B)$. [Suggestion: Begin with $p(\tau) = \tau^2$.]

29. If $c(\tau)$ is the characteristic polynomial of B, show
     that $c(A)$ is singular iff A has an **eigenvalue** in common
     with B.

30. Prove the converse of Lemma 7.1. (Suggestion: Let λ be
     a common eigenvalue of A and B. Explain why A is
     similar to a matrix V whose first column is $[\lambda, 0,\ldots,$
     $0]^{tr}$ and B is similar to matrix W whose first row is
     $[\lambda, 0,\ldots, 0]$; solve $VZ = ZW$ for $Z \neq 0$; use the similarity
     of V to A and B to W to solve $AX = XB$ for X in terms of
     Z, and explain why $X \neq 0$.)

31. If C is any $k \times \ell$ matrix, prove that $AX - XB = C$ has
     one, and only one, solution if A has no eigenvalue in
     common with B. [Suggestion: Let $F(X) = AX - XB$ for
     all $k \times \ell$ matrices X, and explain why F is a nonsingular
     operator on the space of all $k \times \ell$ matrices.]

32. (We'll meet this result when we study differential
     equations later on.) If the real part of every
     eigenvalue of a real matrix A is negative, show that
     there is a *symmetric* (that is, $X = X^{tr}$) matrix X
     satisfying $AX + XA^{tr} = -I$.

33. (We'll meet this when we study the matrix logarithm.)
     Suppose $J = J_r(\lambda) \oplus J_s(\mu)$ and $\lambda \neq \mu$. If $SJ = JS$, show
     that $S = S_1 \oplus S_2$, where $S_1 J_r = J_r S_1$ and $S_2 J_s = J_s S_2$.

Evaluating matrix polynomials. Suppose a $k \times k$ matrix
A and a polynomial $g(\tau) = \sum_{i=0}^{n} \alpha_i \tau^i$ are given ($\alpha_n \neq 0$), and
we want to evaluate $g(A)$ ($= \alpha_0 I + \alpha_1 A + \ldots + \alpha_n A^n$). If n
is large, then this may be difficult to do directly, but if
k is small enough, we can use the Cayley-Hamilton theorem
to find $g(A)$. (You'll see that this special method is more
efficient than using the similarity method as we do not
need to find the matrix P or a Jordan matrix similar to A.)

     Let $c(\tau)$ be the characteristic polynomial of A. If
we divided g by c we would obtain a quotient polynomial $q(\tau)$
and a remainder $r(\tau)$ such that for all τ,

$$g(\tau) = c(\tau)q(\tau) + r(\tau) \qquad (1)$$

where $r(\tau) = \beta_0 + \beta_1\tau + \cdots + \beta_{k-1}\tau^{k-1}$. The degree of $r(\tau)$ is no larger than $k - 1$ because the degree of the divisor $c(\tau)$ is k. Therefore, by the Cayley-Hamilton theorem,

$$g(A) = r(A) \quad (= \beta_0 I + \beta_1 A + \cdots + \beta_{k-1}A^{k-1}) \qquad (2)$$

If k is small compared to n, it may be as awkward to calculate the β_i directly as it was to evaluate $g(A)$ directly, so we need an alternative method for finding the β_i. Suppose the eigenvalues of A are $\lambda_1, \lambda_2, \ldots, \lambda_k$. Letting $\tau = \lambda_i$ in (1) we obtain:

$$\sum_{j=0}^{k-1} \beta_j \lambda_i^j = g(\lambda_i) \quad \text{for} \quad i = 1, 2, \ldots, k \qquad (3)$$

because $c(\lambda_i) = 0$. If there are k distinct eigenvalues for A, then Eq. (3) will determine the β_j.

   Example: As in Sec. 4, let $A = \begin{bmatrix} 1 & 4 \\ 3 & 2 \end{bmatrix}$ and $g(\tau) = \tau^{1010}$. Equation (3) becomes

$$\beta_0 + 5\beta_1 = 5^{1010}$$
$$\beta_0 - 2\beta_1 = (-2)^{1010} \qquad \text{as } c(\tau) = (\tau - 5)(\tau + 2)$$

Solving Eq. (3) we obtain $\beta_0 = \frac{1}{7}\left[2(5^{1010}) + 5(2^{1010})\right]$ and $\beta_1 = \frac{1}{7}\left[5^{1010} - 2^{1010}\right]$. Therefore, by (2),

$$A^{1010} = \beta_0 I + \beta_1 A$$

$$= \frac{1}{7}\begin{bmatrix} 3(5^{1010}) + 2^{1012} & 4(5^{1010}) - 2^{1012} \\ 3(5^{1010}) + 3(2^{1010}) & 4(5^{1010}) + 3(2^{1010}) \end{bmatrix}$$

as we found in Sec. 4.

If there are fewer than k distinct eigenvalues, then we can use (3) and the derivatives of g to obtain enough equations to determine the β_i.

Example: Suppose $A = \begin{bmatrix} 1 & 1 & -1 \\ 1 & 1 & 1 \\ 0 & -1 & 2 \end{bmatrix}$ and we want to

evaluate $A^{100} + 2A^{50}$. We have $c(\tau) = (\tau - 1)^2(\tau - 2)$ and $g(\tau) = \tau^{100} + 2\tau^{50}$. We need to determine the coefficients of $r(\tau) = \beta_0 + \beta_1\tau + \beta_2\tau^2$, the remainder polynomial obtained when g is divided by c. Equation (3) tells us that

$$\begin{aligned} \beta_0 + \beta_1 + \beta_2 &= 3 \\ \beta_0 + \beta_1 + 4\beta_2 &= 2^{100} + 2^{51} \end{aligned} \qquad (4)$$

which isn't enough information to determine the β_i, but if we differentiate Eq. (1) with respect to τ, we obtain $g'(\tau) = c'(\tau)q(\tau) + c(\tau)q'(\tau) + r'(\tau)$ for all τ, and hence

$$g'(\tau) = [2(\tau-1)(\tau-2)+(\tau-1)^2]q(\tau) + c(\tau)q'(\tau) + \beta_1 \\ + 2\beta_2\tau \quad \text{for all } \tau \qquad (5)$$

Letting $\tau = 1$ in Eq. (5), we obtain $g'(1) = \beta_1 + 2\beta_2$, giving us our third equation: $\beta_1 + 2\beta_2 = 200$, which together with Eq. (4) enables us to determine the β_i. Thus

$$A^{100} + 2A^{50} = \beta_0 I + \beta_1 A + \beta_2 A^2$$

where $\beta_0 = 2^{100} + 2^{51} - 400$, $\beta_1 = 606 - 2^{101} - 2^{52}$, $\beta_2 = 2^{100} + 2^{51} - 203$.

We will return to this topic in a somewhat more general way in Sec. 16.

Exercises

34. Using the Cayley-Hamilton theorem:

    a. Evaluate $A^{100} + 3A^{23} + A^{20}$, if $A = \begin{bmatrix} 0 & 1 \\ -1 & 2 \end{bmatrix}$.

    b. Evaluate $\begin{bmatrix} 2 & 2 & 1 \\ 1 & 3 & 1 \\ 1 & 2 & 2 \end{bmatrix}^{1000}$. [*Note:* $c(\tau) = (\tau - 1)^2(\tau - 5)$.]

    c. Evaluate $\begin{bmatrix} 1 & -4 & -1 & -4 \\ 2 & 0 & 5 & -4 \\ -1 & 1 & -2 & 3 \\ -1 & 4 & -1 & 6 \end{bmatrix}^{1000}$. [*Note:* $c(\tau) =$

        $= (\tau - 1)^3(\tau - 2)$.]

35. Do Exercise 34(a) using the similarity method. Is there
    really much saving in time between the methods? How do
    you suppose the efficiency of the methods would compare
    on 34(b)? on (c)?

8. THE QUESTION OF UNIQUENESS OF THE JORDAN FORM

Any Jordan matrix J similar to A is called a *Jordan form* of
A. This raises the question: How many Jordan forms does A
have? This is equivalent to asking, how many Jordan forms
does a Jordan matrix have? The answer was also provided by
Camille Jordan [For a proof see e.g., Hoffman and Kunze
(1961) Sec. 7.3].

   Theorem 1. The only Jordan forms of $J = \overset{m}{\underset{i=1}{\oplus}} J_{n_i}(\lambda_i)$ are
those (Jordan) matrices obtained from J by permuting the m
blocks $J_{n_i}(\lambda_i)$.

   The theorem explains, among other things, why the Jordan
forms of a diagonable matrix are diagonal matrices. Here
is a numerical example of the theorem:

Example: If $A = \begin{bmatrix} 2 & 1 & 0 & 0 & 0 \\ 0 & 2 & 0 & 0 & 0 \\ 0 & 0 & 3 & 0 & 0 \\ 0 & 0 & 0 & 4 & 1 \\ 0 & 0 & 0 & 0 & 4 \end{bmatrix}$, then the following is a

complete list of the Jordan forms of A:

$$J_1 = J_2(2) \oplus 3 \oplus J_2(4)$$

$$J_2 = 3 \oplus J_2(4) \oplus J_2(2)$$

$$J_3 = J_2(4) \oplus J_2(2) \oplus 3$$

$$J_4 = J_2(2) \oplus J_2(4) \oplus 3$$

$$J_5 = 3 \oplus J_2(2) \oplus J_2(4)$$

$$J_6 = J_2(4) \oplus 3 \oplus J_2(2)$$

Exercise

36. Prove directly that $J_1 \sim J_2$ in the example above by finding a matrix P such that $J_1 = PJ_2P^{-1}$.

It is customary, though ambiguous, to speak of *the* Jordan form of A as if there were only one. The excuse for this is that even though there are more than one, they can all be described easily once you know one of them. You may have encountered a similar situation when you heard people speak of *the* solution to a differential equation such as dy/dx = x.

If A is similar to B, then A and B have the same Jordan form (because if A ∼ J and B ∼ A, then B ∼ J); more precisely, they have the same Jordan forms. Conversely, if A and B have the same Jordan form(s), they must be similar. (Can you see why?) Therefore, the seemingly difficult problem of

setting up a procedure for determining whether a given
arbitrary pair of matrices are similar is easily settled
(in principle): Compare their Jordan forms; then A ~ B if
and only if A and B have a common Jordan form. At any rate
we can reduce the problem of determining whether two matrices
are similar to that of finding the Jordan forms of the
matrices.

9. FINDING THE JORDAN FORM

We never have to actually calculate a Jordan form for a
specific matrix in all of the work ahead of us, not even in
the computational problems, because the methods we obtain
involve only the implications of the Jordan form (for example,
we used the Jordan form to prove the Cayley-Hamilton theorem
which in turn enabled us to evaluate matrix polynomials).

Nevertheless, there are several ways to calculate the
Jordan form should you have to do so. One method will be
presented here. For another see, e.g., Gantmacher (1959,
vol. I, Chap. VI), where you can also find a method of
determining a matrix P such that PAP^{-1} is a Jordan form of
A. [*Warning:* No matter which method you choose to find the
Jordan form it will be very sensitive to round-off errors.
Unless all of your calculations are exact (including the
ones used to find the eigenvalues), you may very well get
the wrong Jordan form. This is discussed fully in Franklin
(1968, pp. 232-235).]

A Recipe for Finding the Jordan Form of a k × k Matrix A

(To use this, you will need to recall the idea of the

rank of a matrix [see e.g., Hoffman and Kunze (1961, p.105)] and how to calculate it [see e.g., Lipshutz (1968, pp.90-92)].)

1. Find all the eigenvalues of A, e.g., by finding the roots of the characteristic or minimal polynomial.

2. For each eigenvalue λ and each $1 \leq j \leq k$, find $r_j(\lambda)$, the rank of $(A - \lambda I)^j$. [As soon as you find that $r_{j_0}(\lambda) = r_{j_0+1}(\lambda)$, then $r_j(\lambda) = r_{j_0}(\lambda)$ for all $j \geq j_0$ (see Exercise 38).]

3. For each eigenvalue λ, let

$$b_1(\lambda) = k - 2r_1(\lambda) + r_2(\lambda)$$

   and

$$b_m(\lambda) = r_{m+1}(\lambda) - 2r_m(\lambda) + r_{m-1} \quad \text{for } m \geq 2$$

   Then the Jordan form of A is a direct sum of exactly $b_m(\lambda)$ Jordan blocks of order m for each m and each distinct eigenvalue λ of A.

Here is why the recipe works: Once we know all the eigenvalues of A, the Jordan form is determined when we know the number of Jordan blocks of each order m for each eigenvalue λ. We will therefore calculate the $b_m(\lambda)$.

Recall that the *nullspace* of a k × k matrix M is the set of all vectors x such that Mx = 0 and that the nullspace of M is a vector space. Its dimension is called the *nullity* of M and is related to the rank of M by

$$\text{rank}(M) + \text{nullity}(M) = k \qquad (6)$$

[You will find this in Lipschutz (1968, pp. 126-128, 136-138) and in Hoffman and Kunze (1961, pp. 65-66).]

If a square matrix is *upper triangular* (i.e., all entries below the diagonal are zero), then its nullity is the number of zero columns (e.g., if $U_n = J_n(0)$, then the nullity of $U_n^j = j$ for $1 \leq j \leq n$). For example,

$$U_4 = \begin{bmatrix} 0 & 1 & 0 & 0 \\ 0 & 0 & 1 & 0 \\ 0 & 0 & 0 & 1 \\ 0 & 0 & 0 & 0 \end{bmatrix} \qquad U_4^2 = \begin{bmatrix} 0 & 0 & 1 & 0 \\ 0 & 0 & 0 & 1 \\ 0 & 0 & 0 & 0 \\ 0 & 0 & 0 & 0 \end{bmatrix} \qquad U_4^3 = \begin{bmatrix} 0 & 0 & 0 & 1 \\ 0 & 0 & 0 & 0 \\ 0 & 0 & 0 & 0 \\ 0 & 0 & 0 & 0 \end{bmatrix}$$

$$U_4^{4+j} = 0 \quad \text{for all } j$$

Let λ be an eigenvalue of A and let n_j and r_j be the nullity and rank, respectively, of $(A - \lambda I)^j$ for $j = 1, 2, \ldots, k$. Since rank is similarity invariant (see Hoffman and Kunze (1961, p. 105)), nullity must also be similarity invariant by Eq. (6), and it follows that n_j and r_j are the nullity and rank of $(J - \lambda I)^j$ if J is a Jordan form of A. If $J_m(\mu)$ is a typical Jordan block of J, then $(J - \lambda I)^j$ is a direct sum of blocks $J_m^j(\mu - \lambda) = [(\mu - \lambda)I_m + U_m]^j$. Now the nullity of this direct sum, being the number of zero columns in it, is the sum of the nullities of the blocks. But if $\mu \neq \lambda$, then $J_m^j(\mu - \lambda)$ has no zero columns, so its nullity is zero; therefore, n_j is the sum of the nullities of all the blocks having $\mu = \lambda$. However $B_m^j(\lambda - \mu) = U_m^j$ when $\lambda = \mu$ and

$$\text{nullity}(U_m^j) = \begin{cases} j & \text{when } m \geq j \\ m & \text{when } m < j \end{cases}$$

Therefore, letting $b_m = b_m(\lambda)$, we have

$$n_j = \sum_{m \geq j} jb_m + \sum_{m < j} b_m m \quad \text{for } j \geq 1$$

In particular, $n_1 = \sum_{m \geq 1} b_m$ is the number of Jordan blocks

for the eigenvalue λ, and hence $n_j = j(n_1 - \sum_{m<j} b_m) + \sum_{m<j} mb_m$.

Therefore

$$n_j = jn_1 - \sum_{m<j} (j-m)b_m$$

from which it follows that

$$n_{j+1} - n_j = n_1 - (\sum_{m<j} [(j+1-m) - (j-m)]b_m) - b_j$$

and hence

$$b_j = n_1 + n_j - n_{j+1} - \sum_{m<j} b_m \quad \text{for all } j \geq 1$$

Hence

$$\sum_{m \leq j} b_m = n_1 + n_j - n_{j+1} \quad \text{for all } j$$

Therefore

$$\sum_{m<j} b_m = n_1 + n_{j-1} - n_j$$

substituting in the line before last we get

$$b_1 = 2n_1 - n_2$$
$$b_j = 2n_j - n_{j+1} - n_{j-1} \quad \text{for } j \geq 2$$

If we rewrite this using the fact that $n_j = k - r_j$, we get

$$b_1 = k - 2r_1 + r_2$$
$$b_j = r_{j+1} - 2r_j + r_{j-1} \quad \text{for } j \geq 2$$

Example: Find a Jordan form of

$$A = \begin{bmatrix} 0 & 0 & 0 & 0 & 0 & 1 \\ 2 & 1 & -1 & -1 & 0 & -1 \\ 0 & 0 & 2 & 1 & 0 & 0 \\ 0 & 0 & 0 & 2 & 0 & 0 \\ 0 & 0 & 0 & 0 & 2 & 0 \\ -1 & 0 & 0 & 0 & 0 & 2 \end{bmatrix}$$

$c_A(\tau) = (\tau - 1)^3 (\tau - 2)^3$ is the characteristic polynomial of A.

$$r_1(1) = \text{rank}(A - I) = \text{rank} \begin{bmatrix} -1 & 0 & 0 & 0 & 0 & 1 \\ 2 & 0 & -1 & -1 & 0 & -1 \\ 0 & 0 & 1 & 1 & 0 & 0 \\ 0 & 0 & 0 & 1 & 0 & 0 \\ 0 & 0 & 0 & 0 & 1 & 0 \\ -1 & 0 & 0 & 0 & 0 & 1 \end{bmatrix}$$

$$= \text{rank} \begin{bmatrix} 1 & 0 & 0 & 0 & 0 & -1 \\ 0 & 0 & -1 & -1 & 0 & 1 \\ 0 & 0 & 1 & 1 & 0 & 0 \\ 0 & 0 & 0 & 1 & 0 & 0 \\ 0 & 0 & 0 & 0 & 1 & 0 \\ 0 & 0 & 0 & 0 & 0 & 0 \end{bmatrix}$$

$$= \text{rank} \begin{bmatrix} 1 & 0 & 0 & 0 & 0 & -1 \\ 0 & 0 & 1 & 1 & 0 & 0 \\ 0 & 0 & 0 & 1 & 0 & 0 \\ 0 & 0 & 0 & 0 & 1 & 0 \\ 0 & 0 & 0 & 0 & 0 & 1 \\ 0 & 0 & 0 & 0 & 0 & 0 \end{bmatrix} = 5$$

$$r_2(1) = \text{rank}(A - I)^2 = \text{rank} \begin{bmatrix} 0 & 0 & 0 & 0 & 0 & 0 \\ -1 & 0 & -1 & -2 & 0 & 1 \\ 0 & 0 & 1 & 2 & 0 & 0 \\ 0 & 0 & 0 & 1 & 0 & 0 \\ 0 & 0 & 0 & 0 & 1 & 0 \\ 0 & 0 & 0 & 0 & 0 & 0 \end{bmatrix}$$

$$= \text{rank} \begin{bmatrix} 1 & 0 & 1 & 2 & 0 & -1 \\ 0 & 0 & 1 & 2 & 0 & 0 \\ 0 & 0 & 0 & 1 & 0 & 0 \\ 0 & 0 & 0 & 0 & 1 & 0 \\ 0 & 0 & 0 & 0 & 0 & 0 \\ 0 & 0 & 0 & 0 & 0 & 0 \end{bmatrix} = 4$$

$$r_3(1) = \text{rank}(A - I)^3 = \text{rank} \begin{bmatrix} 0 & 0 & 0 & 0 & 0 & 0 \\ 0 & 0 & -1 & -3 & 0 & 0 \\ 0 & 0 & 1 & 3 & 0 & 0 \\ 0 & 0 & 0 & 1 & 0 & 0 \\ 0 & 0 & 0 & 0 & 1 & 0 \\ 0 & 0 & 0 & 0 & 0 & 0 \end{bmatrix}$$

$$= \text{rank} \begin{bmatrix} 0 & 0 & 1 & 3 & 0 & 0 \\ 0 & 0 & 0 & 1 & 0 & 0 \\ 0 & 0 & 0 & 0 & 1 & 0 \\ 0 & 0 & 0 & 0 & 0 & 0 \\ 0 & 0 & 0 & 0 & 0 & 0 \\ 0 & 0 & 0 & 0 & 0 & 0 \end{bmatrix} = 3$$

$$r_4(1) = \text{rank}(A - I)^4 = \text{rank} \begin{bmatrix} 0 & 0 & 0 & 0 & 0 & 0 \\ 0 & 0 & -1 & -4 & 0 & 0 \\ 0 & 0 & 1 & 4 & 0 & 0 \\ 0 & 0 & 0 & 1 & 0 & 0 \\ 0 & 0 & 0 & 0 & 1 & 0 \\ 0 & 0 & 0 & 0 & 0 & 0 \end{bmatrix}$$

$$= \text{rank} \begin{bmatrix} 0 & 0 & 1 & 4 & 0 & 0 \\ 0 & 0 & 0 & 1 & 0 & 0 \\ 0 & 0 & 0 & 0 & 1 & 0 \\ 0 & 0 & 0 & 0 & 0 & 0 \\ 0 & 0 & 0 & 0 & 0 & 0 \\ 0 & 0 & 0 & 0 & 0 & 0 \end{bmatrix} = 3$$

$$b_1(1) = 6 - 2(5) + 4 = 0$$

$$b_2(1) = 3 - 2(4) + 5 = 0$$

$$b_3(1) = 3 - 2(3) + 4 = 1$$

$$b_j(1) = 3 - 2(3) + 3 = 0 \quad \text{for all } j \geq 4$$

$$r_1(2) = \text{rank}(A - 2I) = \text{rank} \begin{bmatrix} -2 & 0 & 0 & 0 & 0 & 1 \\ 2 & -1 & -1 & -1 & 0 & -1 \\ 0 & 0 & 0 & 0 & 1 & 0 \\ 0 & 0 & 0 & 0 & 0 & 0 \\ -1 & 0 & 0 & 0 & 0 & 0 \end{bmatrix}$$

$$= \text{rank} \begin{bmatrix} 1 & 0 & 0 & 0 & 0 & 0 \\ 0 & 1 & 1 & 1 & 0 & 1 \\ 0 & 0 & 0 & 1 & 0 & 0 \\ 0 & 0 & 0 & 0 & 0 & 1 \\ 0 & 0 & 0 & 0 & 0 & 0 \\ 0 & 0 & 0 & 0 & 0 & 0 \end{bmatrix} = 4$$

$$r_2(2) = \text{rank}(A - 2I)^2 = \text{rank} \begin{bmatrix} 3 & 0 & 0 & 0 & 0 & -2 \\ -3 & 1 & 1 & 0 & 0 & 3 \\ 0 & 0 & 0 & 0 & 0 & 0 \\ 0 & 0 & 0 & 0 & 0 & 0 \\ 0 & 0 & 0 & 0 & 0 & 0 \\ 2 & 0 & 0 & 0 & 0 & 1 \end{bmatrix}$$

$$= \text{rank} \begin{bmatrix} 1 & 0 & 0 & 0 & 0 & -3 \\ 0 & 1 & 1 & 0 & 0 & 1 \\ 0 & 0 & 0 & 0 & 0 & 1 \\ 0 & 0 & 0 & 0 & 0 & 0 \\ 0 & 0 & 0 & 0 & 0 & 0 \\ 0 & 0 & 0 & 0 & 0 & 0 \end{bmatrix} = 3$$

$$r_3(2) = \text{rank}(A - 2I)^3 = \text{rank} \begin{bmatrix} -4 & 0 & 0 & 0 & 0 & 5 \\ 7 & -1 & -1 & 0 & 0 & -8 \\ 0 & 0 & 0 & 0 & 0 & 0 \\ 0 & 0 & 0 & 0 & 0 & 0 \\ 0 & 0 & 0 & 0 & 0 & 0 \\ -3 & 0 & 0 & 0 & 0 & 2 \end{bmatrix}$$

$$= \text{rank} \begin{bmatrix} -4 & 0 & 5 & 0 & 0 & 0 \\ 7 & 1 & -8 & 0 & 0 & 0 \\ 0 & 0 & 0 & 0 & 0 & 0 \\ 0 & 0 & 0 & 0 & 0 & 0 \\ 0 & 0 & 0 & 0 & 0 & 0 \\ -3 & 0 & 2 & 0 & 0 & 0 \end{bmatrix} = 3$$

$b_1(2) = 6 - 2(4) + 3 = 1$

$b_2(2) = 3 - 2(3) + 4 = 1$

$b_j(2) = 3 - 2(3) + 3 = 0$ all $j \geq 3$

Therefore J has one 3 × 3 block corresponding to 1, one 1 × 1
block corresponding to 2, and one 2 × 2 block corresponding
to 2; so

$$J = \begin{bmatrix} 1 & 1 & 0 & 0 & 0 & 0 \\ 0 & 1 & 1 & 0 & 0 & 0 \\ 0 & 0 & 1 & 0 & 0 & 0 \\ 0 & 0 & 0 & 2 & 1 & 0 \\ 0 & 0 & 0 & 0 & 2 & 0 \\ 0 & 0 & 0 & 0 & 0 & 2 \end{bmatrix}$$

Example: Find a Jordan form of $A = \begin{bmatrix} 3 & 1 \\ 2 & 4 \end{bmatrix}$. The
characteristic polynomial is $c_A(\tau) = (\tau - 5)(\tau - 2)$; since

the Jordan form J has the same two distinct eigenvalues 2

and 5, it must be that $J = \begin{bmatrix} 2 & 0 \\ 0 & 5 \end{bmatrix} \left(\text{or } \begin{bmatrix} 5 & 0 \\ 0 & 2 \end{bmatrix} \right)$. Of course we

could use the recipe to find that

$$r_1(2) = \text{rank} \begin{bmatrix} 1 & 1 \\ 2 & 2 \end{bmatrix} = 1 \qquad r_2(2) = \text{rank} \begin{bmatrix} 3 & 3 \\ 6 & 6 \end{bmatrix} = 1$$

$$r_1(5) = \text{rank} \begin{bmatrix} -2 & 1 \\ 2 & -1 \end{bmatrix} = 1 \qquad r_2(5) = \text{rank} \begin{bmatrix} 6 & -3 \\ -6 & 3 \end{bmatrix} = 1$$

Hence $b_1(2) = 2 - 2 + 1 = 1$ and $b_2(5) = 2 - 2 + 1 = 1$; so

$J = \begin{bmatrix} 2 & 0 \\ 0 & 5 \end{bmatrix}$ is one Jordan form.

Example: Find a Jordan form of $A = \begin{bmatrix} 2 & -1 \\ 1 & 4 \end{bmatrix}$; $c_A(\tau) =$

$(\tau - 3)^2$. Since a Jordan form of A has the same

characteristic polynomial, either $J = \begin{bmatrix} 3 & 0 \\ 0 & 3 \end{bmatrix}$ or $J = \begin{bmatrix} 3 & 1 \\ 0 & 3 \end{bmatrix}$.

Since A isn't similar to 3I (see Exercise 2), it must be

that $\begin{bmatrix} 3 & 1 \\ 0 & 3 \end{bmatrix}$ is the Jordan form of A. Of course we could

use the recipe to find that

$$r_1(3) = \text{rank} \begin{bmatrix} -1 & -1 \\ 1 & 1 \end{bmatrix} = 1$$

$$r_2(3) = \text{rank} \begin{matrix} 0 & 0 \\ 0 & 0 \end{matrix} = 0$$

so $r_j(3) = 0$ for all $j \geq 2$. Hence $b_1(3) = 2 - 2 + 0 = 0$

and $b_2(3) = r_3(3) - 2r_2(3) + r_1(3) = 1$. Therefore $J = \begin{bmatrix} 3 & 1 \\ 0 & 3 \end{bmatrix}$.

Example: Find a Jordan form of

$$A = \begin{bmatrix} 1 & 1 & 1 & 1 \\ 1 & 1 & 1 & 1 \\ 1 & 1 & 1 & 1 \\ 1 & 1 & 1 & 1 \end{bmatrix}$$

Notice that the rank of A is 1 so a Jordan form of A must

have rank 1 too, therefore J has the form

$$J = \begin{bmatrix} \lambda & 0 & 0 & 0 \\ 0 & 0 & 0 & 0 \\ 0 & 0 & 0 & 0 \\ 0 & 0 & 0 & 0 \end{bmatrix}$$

Now $A[1, 1, 1, 1]^{tr} = 4[1, 1, 1, 1]^{tr}$, so 4 is an eigenvalue

of A, and hence $J = \begin{bmatrix} 4 & 0 & 0 & 0 \\ 0 & 0 & 0 & 0 \\ 0 & 0 & 0 & 0 \\ 0 & 0 & 0 & 0 \end{bmatrix}$. Of course we could use

the recipe:

We first find that $c_A(\tau) = \tau^3(\tau - 4)$. Then

$$r_1(0) = \text{rank } A = 1$$
$$r_2(0) = \text{rank } A^2 = \text{rank } 4A = \text{rank } A = 1$$
$$r_j(0) = \text{rank } 4^{j-1}A = \text{rank } A = 1$$

Therefore

$$b_1(0) = 4 - 2 + 1 = 3$$
$$b_j(0) = 1 - 2 + 1 = 0 \quad \text{for all } j > 1$$

Since J has three 1×1 blocks of 0, it must have one 1×1

block of 4. Therefore

$$J = \begin{bmatrix} 4 & 0 & 0 & 0 \\ 0 & 0 & 0 & 0 \\ 0 & 0 & 0 & 0 \\ 0 & 0 & 0 & 0 \end{bmatrix}$$

If we had started trying to calculate $r_j(4)$, we would have

found $r_1(4) = 3$, $r_2(4) = 3$, $r_j(4) = 3$ for all j so that

$$b_1(4) = 4 - 6 + 3 = 1$$
$$b_j(4) = 3 - 6 + 3 = 0 \quad \text{for all } j > 1$$

and hence J has exactly one 1×1 block for the eigenvalue 4.

Exercises

37. If λ is an eigenvalue of A, then A - λI is similar to a
     direct sum of Jordan blocks that look like

$$S \equiv J_{p_1}(0) \oplus J_{p_2} \oplus \cdots \oplus J_{p_t}(0) \oplus \overset{m}{\underset{i=t+1}{\oplus}} J_{p_i}(\lambda_i - \lambda)$$

where $\lambda_i \neq \lambda$ for i > t. Express the nullity of S (and
hence of A - λI) in terms of one of the parameters in
the equation defining S.

38. If r_j = rank(A - λI)^j, show that if $r_{j_0} = r_{j_0+1}$, then
     $r_j = r_{j_0}$ for all j $\geq$ j_0. (*Hint:*

     1. $(A - \lambda I)^j \sim J_{p_1}^i(0) \oplus J_{p_2}^i(0) \oplus \cdots \oplus J_{p_t}^j(0) \oplus$

$$\overset{m}{\underset{i=t+1}{\oplus}} J_{p_i}^j(\lambda_i - \lambda)$$

     2. nullity$(J_p^j(0)) = \begin{cases} j & \text{if} \quad j \leq p \\ p & \text{if} \quad j > p \end{cases}$

     3. nullity (A - λI)^j = ?

39. If $\tau(\tau - 1)^2$ is the minimal polynomial of a 4 × 4 matrix
     A, what are the possible Jordan forms of A?

40. Find a Jordan form of each of the following matrices:

     a. $\begin{bmatrix} 0 & 1 \\ -1 & 2 \end{bmatrix}$

     b. $\begin{bmatrix} 2 & 2 & 1 \\ 1 & 3 & 1 \\ 1 & 2 & 2 \end{bmatrix}$ [*Note:* c(τ) = $(\tau - 1)^2(\tau - 5)$]

     c. $\begin{bmatrix} 2 & 0 & 3 & 1 \\ 0 & 2 & 1 & 2 \\ 0 & 0 & 1 & 2 \\ 0 & 0 & 0 & 1 \end{bmatrix}$

     d. $\begin{bmatrix} 1 & -4 & -1 & -4 \\ 2 & 0 & 5 & -4 \\ -1 & 1 & -2 & 3 \\ -1 & 4 & -1 & 6 \end{bmatrix}$ [*Note:* c(τ) = $(\tau - 1)^3(\tau - 2)$]

e. $\begin{bmatrix} 2 & 2 & 3 & 1 \\ 0 & 2 & 1 & 2 \\ 0 & 0 & 1 & 2 \\ 0 & 0 & 0 & 1 \end{bmatrix}$

10. SEQUENCES OF MATRICES

Suppose M_1, M_2, ..., M_n, ... is an infinite sequence of $k \times \ell$ matrices. Letting $m_{ij}^{(n)}$ denote the ij^{th} entry in M_n, we say that a $k \times \ell$ matrix M is the *limit* of the given sequence (written $M = \lim_{n \to \infty} M_n$) iff for each i, j we have $m_{ij} = \lim_{n \to \infty} m_{ij}^{(n)}$. For example,

$$\begin{bmatrix} 1 \\ 2 \\ 3 \end{bmatrix} = \lim_{n \to \infty} \begin{bmatrix} 1 - \frac{1}{n} \\ 2 + \frac{1}{n} \\ 3 \end{bmatrix} \text{ and } \begin{bmatrix} e & 1 \\ 0 & 1 \end{bmatrix} = \lim_{n \to \infty} \begin{bmatrix} \left(1 + \frac{1}{n}\right)^n & \frac{n+1}{n} \\ \frac{2^n}{n!} & 1 \end{bmatrix}$$

If a sequence of matrices has a limit, we say that it is a *convergent* sequence.

Proposition. If $\lim_{n \to \infty} M_n = M$ and $\lim_{n \to \infty} L_n = L$, all M_n are $s \times p$ and all L_n are $p \times m$, then $\lim_{n \to \infty} M_n L_n = ML$.

Proof: Let $K_n = M_n L_n$ and $K = ML$; then $\ell_{ij}^{(n)} = \sum_{t=1}^{p} m_{it}^{(n)} k_{tj}^{(n)}$ and $k_{ij} = \sum_{t=1}^{p} m_{it} \ell_{tj}$, but $\lim_{n \to \infty} m_{it}^{(n)} = m_{it}$ and $\lim_{n \to \infty} \ell_{tj}^{(n)} = \ell_{tj}$ for all i, t, j, and so $\lim_{n \to \infty} m_{it}^{(n)} \ell_{tj}^{(n)} = m_{it} \ell_{tj}$. Hence

$$\lim_{n \to \infty} k_{ij}^{(n)} = \lim_{n \to \infty} \sum_{t=1}^{p} m_{it}^{(n)} \ell_{tj}^{(n)} = \sum_{t=1}^{p} \lim_{n \to \infty} m_{it}^{(n)} \ell_{tj}^{(n)}$$

$$= \sum_{t=1}^{p} m_{it} \ell_{tj} = k_{ij} \text{ for all i, j}$$

Therefore, $\lim_{n \to \infty} M_n L_n = ML$.   []

Exercises

41. Suppose $\lim_{n\to\infty} M_n = M$ and α is any scalar. Show that

    $\lim_{n\to\infty} \alpha M_n = \alpha M$.

42. Find an example of a sequence of 2×2 invertible matrices having a noninvertible limit.

43. Suppose A, A_1, A_2, ..., A_n, ... and C_1, C_2, ..., C_n, ... are $k \times \ell$ and B, B_1, B_2, ..., B_n, ... are $\ell \times m$.

    a. If $\lim_{n\to\infty} B_n = B$ and $\lim_{n\to\infty} A_n = A$, show that $\lim_{n\to\infty}(A_n + C_n) = A + C$.

    b. If $\lim_{n\to\infty} A_n B_n$ exists, must $\lim_{n\to\infty} A_n$ and $\lim_{n\to\infty} B_n$ necessarily exist? If so, prove it. If not, provide a counterexample.

    c. If $\lim_{n\to\infty} A_n = A$, show that $\lim_{n\to\infty} (A_n B) = AB$.

44. If A_1, A_2, ..., A_n, ... are $k \times k$, P is a $k \times k$ invertible matrix, and A_1, A_2, ... converges, show that $PA_1 P^{-1}$, ..., $PA_n P^{-1}$, ... converges and $\lim_{n\to\infty} PA_n P^{-1} = P(\lim_{n\to\infty} A_n)P^{-1}$.

11. POWERS OF MATRICES

Suppose A is a $k \times k$ matrix. For the work that follows we need to investigate when the powers A, A^2, A^3, ..., A^n, ... are a convergent sequence. We say that A is *power convergent* when this occurs.

Exercises

45. Show that power-convergence is a similarity invariant.

46. Show that $\displaystyle\bigoplus_{i=1}^{m} A_i$ is power convergent iff each A_i is

power convergent and that $\displaystyle\lim_{n\to\infty}\left(\bigoplus_{i=1}^{m} A_i\right)^n = \bigoplus_{i=1}^{m} \lim_{n\to\infty} A_i^n$.

As a result of these two exercises and Jordan's theorem, we see that we may restrict our attention to Jordan blocks.

Let $U = J_\ell(0)$ and $I = I_\ell$ (the $\ell \times \ell$ identity matrix), then $J_\ell(\lambda) = \lambda I + U$; so by the binomial theorem [see Exercise 20(c)], $[J_\ell(\lambda)]^n = \lambda^n I + n\lambda^{n-1}U + \dfrac{n(n-1)}{2}\lambda^{n-2}U^2 +$

$+\cdots+ \binom{n}{j}\lambda^{n-j}U^j + \cdots + U^n$. Notice that for $n \geq \ell$, the sum

ends with the term $\binom{n}{\ell-1}\lambda^{n-\ell+1}U^{\ell-1}$ by Exercise 22(a).

Therefore $\left[J_\ell(\lambda)\right]^n = \displaystyle\sum_{j=0}^{\ell-1}\binom{n}{j}\lambda^{n-j}U^j$ for all $n \geq \ell$. That is,

$$[J_\ell(\lambda)]^n = \begin{bmatrix} \lambda^n & \binom{n}{1}\lambda^{n-1} & \binom{n}{2}\lambda^{n-2} & \cdots & \binom{n}{n-\ell+1}\lambda^{n-\ell+1} \\ 0 & \lambda^n & \binom{n}{1}\lambda^{n-1} & \cdots & \binom{n}{n-\ell}\lambda^{n-\ell} \\ 0 & 0 & \lambda^n & \cdots & \cdot \\ \cdot & \cdot & \cdot & & \cdot \\ \cdot & \cdot & \cdot & \lambda^n & \binom{n}{1}\lambda^{n-1} \\ 0 & 0 & 0 & 0 & \lambda^n \end{bmatrix} \quad (7)$$

when $n \geq \ell$.

Exercises

47. Show that $\displaystyle\lim_{n\to\infty} \binom{n}{j}\lambda^{n-j} = 0$ if $|\lambda| < 1$ when $n \geq j$.

48.   If $\lim\limits_{n\to\infty} \lambda^n$ exists and $|\lambda| = 1$, show that $\lambda = 1$.   (*Hint:*

$\lambda \lim\limits_{n\to\infty} \lambda^n = \lim\limits_{n\to\infty} \lambda^n$.)

Lemma 11.1. $J_\ell(\lambda)$ is power convergent iff $|\lambda| < 1$ or $\lambda = 1$ and $\ell = 1$.

Proof: Suppose $J_\ell(\lambda)$ is power convergent.   Then, by the definition of limit of a sequence of matrices and by Eq. (7), we see that $\lim\limits_{n\to\infty} \lambda^n$ exists, and hence $|\lambda| \leq 1$.   If $|\lambda| = 1$, then (Exercise 48) $\lambda = 1$; if $\ell > 1$ also, then Eq. (7) would imply that $\lim\limits_{n\to\infty} n1^{n-1}$ exists which is impossible, and therefore $\ell = 1$.   Recapitulating, we have shown that if $J_\ell(\lambda)$ is power convergent, then $|\lambda| < 1$ or $\ell = \lambda = 1$. Conversely, if $\ell = \lambda = 1$, then $J_\ell(\lambda) = 1$ (= [1] if you insist), and so $\lim\limits_{n\to\infty} [J_\ell(\lambda)]^n$ exists.   If $|\lambda| < 1$, then (Exercise 47) $\lim\limits_{n\to\infty} [J_\ell(\lambda)]^n$ exists and is in fact the $\ell \times \ell$ zero matrix.   []

We have also proven, on the way the following lemma.

Lemma 11.2.   If $K = \lim\limits_{n\to\infty} J_\ell^n(\lambda)$, then (1) $K = 0_\ell$ (the $\ell \times \ell$ zero matrix) if $|\lambda| < 1$, and (2) $J_\ell(\lambda) = J_1(1) = 1 = K$, if $|\lambda| = 1$.

Theorem 2.   A is a power convergent iff for each eigenvalue λ of A,   (1) $|\lambda| \leq 1$ and   (2) if $|\lambda| = 1$, then $\lambda = 1$ and the Jordan blocks corresponding to $\lambda = 1$ in the Jordan form for A are 1×1.

Proof: Let $J = \bigoplus\limits_{i=1}^{m} J_{n_i}(\lambda_i)$ be the Jordan form of A. A

is power convergent iff J is power convergent (Exercise 45)

iff each of its blocks $J_{n_i}(\lambda_i)$ is power convergent (Exercise

46) iff $|\lambda_i| < 1$ or $n_i = \lambda_i = 1$ (for each $i \leq m$) by Lemma

11.1.

Exercises

49. If $J = \bigoplus\limits_{i=1}^{m} J_{n_i}(\lambda_i)$ is power convergent and r is the

rank of $J - I$, show that $\lim\limits_{n\to\infty} J^n$ is a diagonal matrix

whose diagonal entries consist of r zeroes and $k - r$

ones.

50. If $L = \lim\limits_{n\to\infty} A^n$, show that $L \sim I_{k-r} \oplus O_r$, where r is the

rank of $A - I$. (Convention: "$I_0 \oplus O_k$" means O_k and

"$I_k \oplus O_0$" means I_k).

51. Show that $\lim\limits_{n\to\infty} A^n = 0$ iff $|\lambda| < 1$ for all eigenvalues

λ of A.

52. If $L = \lim\limits_{n\to\infty} A^n$, show that L is idempotent.

53. If $L = \lim\limits_{n\to\infty} A^n$, show that $(A - L)^n = A^n - L$.

A matrix A is said to be *power-bounded* iff there is a

$\beta > 0$ such that for every n, i, and j, $\left|a_{ij}^{(n)}\right| < \beta$. Thus

$A = \begin{bmatrix} 0.1 & 0.9 \\ 0.2 & 0.8 \end{bmatrix}$ is power bounded because (as you can show by

induction on n) $0 < a_{ij}^{(n)} < 1$ for all $n > 0$ and all

$i \leq i, j \leq 2$.

Exercises

54. Show that power boundedness is a similarity invariant.

55. Show that $\bigoplus\limits_{i=1}^{m} A_i$ is power bounded iff each A_i is power

bounded.

56. State and prove a theorem about power boundedness

analogous to Theorem 2.

12. SERIES OF MATRICES

If $A_1, A_2, \ldots, A_n, \ldots$ is a sequence of $k \times \ell$ matrices,

then the sequence $A_1, A_1 + A_2, \ldots, \sum\limits_{n=1}^{m} A_n, \ldots$ is called the

infinite series generated by $A_1, A_2, \ldots, A_n, \ldots$. The

series is denoted by $\sum\limits_{n=1}^{\infty} A_n$ or by $A_1 + A_2 + \cdots + A_n + \cdots$

and the terms $\sum\limits_{n=1}^{m} A_n$ are called *partial sums*. If the series

converges, i.e., if $\lim\limits_{m \to \infty} \left(\sum\limits_{n=1}^{m} A_n \right)$ exists, then this limit is

also referred to by the same symbol $\sum\limits_{n=1}^{\infty} A_n$. I would

apologize for this ambiguous notation if it were mine.

Exercises

(*Note*: Matrices in this exercise set are $k \times \ell$.)

57. If $\sum\limits_{n=1}^{\infty} A_n$ converges, prove that $\lim\limits_{n \to \infty} A_n = 0$.

58. Find a counter example to the converse, using a series

of 2×2 matrices.

59. If $\sum\limits_{n=1}^{\infty} A_n = S$ and $\sum\limits_{n=1}^{\infty} B_n = T$, show that

a. $\sum\limits_{n=1}^{\infty} (A_n + B_n)$ converges and $\sum\limits_{n=1}^{\infty} (A_n + B_n) = S + T$

b. $\sum\limits_{n=1}^{\infty} \alpha A_n = \alpha S$ for all α.

c. $\sum\limits_{n=1}^{\infty} (A_n x) = Sx$ for every $\ell \times 1$ vector x

60. Is it true that $\sum\limits_{n=1}^{\infty} A_n = A$ iff $\sum\limits_{n=1}^{\infty} a_{ij}^{(n)} = a_{ij}$ for all

$1 \leq i \leq k$ and $1 < j \leq \ell$?

13. POWER SERIES: THE LAGRANGE-SYLVESTER THEOREM

Given scalars α_0, α_1, ..., α_n, ... and a matrix T, the

series $\sum\limits_{n=0}^{\infty} \alpha_n T^n$ is called a *power series* in T. For example,

in Sec. 5 we encountered the power series $\sum\limits_{n=0}^{\infty} \begin{bmatrix} 0.1 & 0.7 \\ 0.3 & 0.6 \end{bmatrix}^n$

and we observed (in somewhat different language) that

$\sum\limits_{n=0}^{\infty} \begin{bmatrix} 0.1 & 0.7 \\ 0.3 & 0.6 \end{bmatrix}^n = P \begin{bmatrix} (1-\lambda)^{-1} & 0 \\ 0 & (1-\mu)^{-1} \end{bmatrix}^{-1} P^{-1}$ for some P, μ,

and λ.

As with scalar series, given α_0, α_1, ..., α_n, ..., we

will need to know for which matrices T the series $\sum\limits_{n=0}^{\infty} \alpha_n T^n$

will converge. To answer this question we will first discuss

an important connection between the scalar-valued function f

defined by $f(\tau) = \sum\limits_{n=0}^{\infty} \alpha_n \tau^n$ (assuming the series converges for

some $\tau \neq 0$), and the matrix series $\sum\limits_{n=0}^{\infty} \alpha_n T^n$.

Theorem 3 (Lagrange-Sylvester). Suppose[†]

[†] We conform to the usual convention that "$\rho = \infty$," means the
series converges for all τ in which case the hypothesis
(b) is satisfied for all T. See Knopp (1952, pp. 84, 85)
to review the radius of convergence.

a. $\rho > 0$ is the radius of convergence of $\Sigma \alpha_n \tau^n$ and $f(\tau) = \Sigma \alpha_n \tau^n$ for all $|\tau| < \rho$

b. $|\lambda| < \rho$ for all eigenvalues λ of T

c. $f^{(j)}$ denotes the jth derivative of f, $U_n = J_n(0)$ and $T = PJP^{-1}$ where $J = \overset{m}{\underset{i=1}{\oplus}} J_{n_i}(\lambda_i)$ is the Jordan form of T,

    then

$$\sum_{n=0}^{\infty} \alpha_n T^n = \left(P \overset{m}{\underset{i=1}{\oplus}} \sum_{j=0}^{n_i-1} \frac{f^{(j)}(\lambda_i)}{j!} U_{n_i}^j \right) P^{-1} \qquad (8)$$

Example 1: If $T = P(J_4(3) \oplus J_3(1))P^{-1}$, then

$$\sum_{n=0}^{\infty} \alpha_n T^n = P \begin{bmatrix} f(3) & f'(3) & \frac{1}{2}f''(3) & \frac{1}{6}f'''(3) & 0 & 0 & 0 \\ 0 & f(3) & f'(3) & \frac{1}{2}f''(3) & 0 & 0 & 0 \\ 0 & 0 & f(3) & f'(3) & 0 & 0 & 0 \\ 0 & 0 & 0 & f(3) & 0 & 0 & 0 \\ 0 & 0 & 0 & 0 & f(1) & f'(1) & \frac{1}{2}f''(1) \\ 0 & 0 & 0 & 0 & 0 & f(1) & f'(1) \\ 0 & 0 & 0 & 0 & 0 & 0 & f(1) \end{bmatrix} P^{-1}$$

because the block

$$\sum_{j=0}^{3} \frac{f^{(j)}(3)}{j!} U_4^j = f(3)\begin{bmatrix} 1&0&0&0 \\ 0&1&0&0 \\ 0&0&1&0 \\ 0&0&0&1 \end{bmatrix} + f^{(1)}(3)\begin{bmatrix} 0&1&0&0 \\ 0&0&1&0 \\ 0&0&0&1 \\ 0&0&0&0 \end{bmatrix}$$

$$+ \frac{f^{(2)}(3)}{2}\begin{bmatrix} 0&0&1&0 \\ 0&0&0&1 \\ 0&0&0&0 \\ 0&0&0&0 \end{bmatrix} + \frac{f^{(3)}(3)}{3!}\begin{bmatrix} 0&0&0&1 \\ 0&0&0&0 \\ 0&0&0&0 \\ 0&0&0&0 \end{bmatrix}$$

and the block

$$\sum_{j=0}^{2} \frac{f^{(j)}(1)}{j!} U_3^j = f(1)\begin{bmatrix} 1 & 0 & 0 \\ 0 & 1 & 0 \\ 0 & 0 & 1 \end{bmatrix} + f^{(1)}(1)\begin{bmatrix} 0 & 1 & 0 \\ 0 & 0 & 1 \\ 0 & 0 & 0 \end{bmatrix} +$$

$$+ \frac{f^{(2)}(1)}{2}\begin{bmatrix} 0 & 0 & 1 \\ 0 & 0 & 0 \\ 0 & 0 & 0 \end{bmatrix}$$

Example 2:   In particular, if T is as in Example 1 and

$f(\tau) = 1 + \frac{1}{6}\tau + \frac{1}{36}\tau^2 + \cdots + \frac{1}{6^n}\tau^n + \cdots = (1 - \frac{\tau}{6})^{-1}$, then

$f'(\tau) = \frac{1}{6}\left(1 - \frac{\tau}{6}\right)^{-2}$, $f''(\tau) = \frac{1}{18}\left(1 - \frac{\tau}{6}\right)^{-3}$, $f'''(\tau) = \frac{1}{36}\left(1 - \frac{\tau}{6}\right)^{-4}$.

Letting f(T) denote $\sum_{n=0}^{\infty} (\frac{1}{6})^n T^n$, we have

$$f(T) = P\begin{bmatrix} 2 & \frac{2}{3} & \frac{2}{9} & \frac{2}{27} \\ 0 & 2 & \frac{2}{3} & \frac{2}{9} \\ 0 & 0 & 2 & \frac{2}{3} \\ 0 & 0 & 0 & 2 \end{bmatrix} \oplus \begin{bmatrix} \frac{6}{5} & \frac{6}{25} & \frac{6}{125} \\ 0 & \frac{6}{5} & \frac{6}{25} \\ 0 & 0 & \frac{6}{5} \end{bmatrix} P^{-1}$$

You can verify that $f(T)[I - (\frac{1}{6})T] = I$; so $f(T) = [I - (\frac{1}{6})T]^{-1}$.

Example 3:   Suppose $T = J_4(3) \oplus J_3(1)$ and $f(\tau) =$
$2 + \tau + 3^2$ (that is, $\alpha_0 = 2$, $\alpha_1 = 1$, $\alpha_2 = 3$, $\alpha_j = 0$ if
$j > 2$).   Then substituting in the matrix of Example 1 we get

$$2I + T + 3T^2 = \begin{bmatrix} 32 & 19 & 3 & 0 & 0 & 0 & 0 \\ 0 & 32 & 19 & 3 & 0 & 0 & 0 \\ 0 & 0 & 32 & 19 & 0 & 0 & 0 \\ 0 & 0 & 0 & 32 & 0 & 0 & 0 \\ 0 & 0 & 0 & 0 & 6 & 7 & 3 \\ 0 & 0 & 0 & 0 & 0 & 6 & 7 \\ 0 & 0 & 0 & 0 & 0 & 0 & 6 \end{bmatrix}$$

You might compute $2I + T + 3T^2$ directly and compare.

Proof of Theorem 3: Differentiating term by term, as we

may when $|\tau| < \rho$, we have $f^{(j)}(\tau) = \sum_{n=j}^{\infty} \alpha_n \frac{n!}{(n - j)!} \tau^{n-j}$ for all $j \geq 0$. It will be convenient to set $f_\ell^{(j)} =$

$$\sum_{n=j}^{\ell} \alpha_n \frac{n!}{(n-j)!} \tau^{n-j} \text{ for each } \ell \geq 0 \ (f^{(0)} = f).$$

1. $f_\ell(J(\lambda)) = \sum_{n=0}^{\ell} \alpha_n J(\lambda)^n = \sum_{n=0}^{\ell} \alpha_n [\lambda I + U]^n$ where $J(\lambda)$

is an $r \times r$ Jordan block

$J = J_r(0)$ and $I = I_r$ as in

Sec. 11.

$= \sum_{n=0}^{\ell} \alpha_n \sum_{j=0}^{r-1} \binom{n}{j} \lambda^{n-j} U^j$ If $n < r - 1$ then we

could have $j > n$ so

we put $\binom{n}{j} = 0$ when

$j > n$.

$= \sum_{n=0}^{\ell} \sum_{j=0}^{r-1} \alpha_n \binom{n}{j} \lambda^{n-j} U^j$

$= \sum_{j=0}^{r-1} \sum_{n=j}^{\ell} \alpha_n \lambda^{n-j} \frac{n!}{(n-j)!} \frac{1}{j!} U^j$ Since

$\binom{n}{j} = 0$ when $j > n$

and $\binom{n}{j} = \dfrac{n!}{(n-j)!j!}$

when $j \leq n$.

$= \sum_{j=0}^{r-1} \dfrac{f_\ell^{(j)}(\lambda)}{j!} U^j$

Now $T = P\left(\bigoplus_{i=1}^{m} J_{n_i}(\lambda_i)\right) P^{-1}$. Therefore (Exercise 3),

we have

2. $\sum_{n=0}^{\ell} \alpha_n T^n = f_\ell(T) = P\left(f_\ell\left(\bigoplus_{i=1}^{m}(J_{n_i}(\lambda_i))\right)\right) P^{-1}$

Therefore (Exercise 16)

$$= P\left(\bigoplus_{i=1}^{m} f_\ell(J_{n_i}(\lambda_i))\right) P^{-1}$$

$$= P\left(\overset{m}{\underset{i=1}{\oplus}} \overset{n_i-1}{\underset{j=0}{\Sigma}} \frac{f_\ell^{(j)}(\lambda_i)}{j!} U_{n_i}^j\right)P^{-1}$$

for all ℓ by (1). But $\lim_{\ell\to\infty} f_\ell^{(j)}(\tau) = f^{(j)}(\tau)$ for

all j and all $|\tau| < \rho$. Since $|\lambda_i| < \rho$, (b) implies

that $\lim_{\ell\to\infty} \overset{\ell}{\underset{n=0}{\Sigma}} \alpha_n T^n$ exists and equals

$$P\left(\overset{m}{\underset{i=1}{\oplus}} \overset{n_i-1}{\underset{j=1}{\Sigma}} \frac{f^{(j)}(\lambda_i)}{j!} U_{n_i}^j\right)P^{-1}$$

In other words,

$$\overset{\infty}{\underset{m=0}{\Sigma}} \alpha_n T^n = P\left(\overset{m}{\underset{i=1}{\oplus}} \overset{n_i-1}{\underset{j=0}{\Sigma}} \frac{f^{(j)}(\lambda_i)}{j!} U_{n_i}^j\right)P^{-1} \qquad []$$

Corollary. If ρ is the radius of convergence of the
scalar power series $\Sigma \alpha_n \tau^n$ and $|\lambda| < \rho$ for every eigenvalue
λ of T then the matrix power series $\overset{\infty}{\underset{n=0}{\Sigma}} \alpha_n T^n$ converges.

The *spectral radius* $|T|$ of a matrix T is defined by
$|T| = \max\{|\lambda|: \lambda$ is an eigenvalue of T$\}$. Using this
terminology we can restate the corollary:

Corollary. If the radius of convergence of $\Sigma \alpha_n \tau^n$
exceeds the spectral radius of T, then $\Sigma \alpha_n T^n$ converges.

Exercise

61. The series $\Sigma \alpha_n T^n$ diverges if the spectral radius of T
exceeds the radius of convergence of $\Sigma \alpha_n \tau^n$. (Suggestion:
Use Exercise 44.)

Thus the corollary and Exercise 61 give us the matrix analogue of the familiar scalar theorem: $\Sigma \alpha_n \tau^n$ converges for $|\tau| < \rho$ and diverges for $|\tau| > \rho$.

Theorem 4. If ρ is the radius of convergence of $\Sigma \alpha_n \tau^n$ and T is any matrix then the series $\Sigma \alpha_n T^n$ converges if $|T| < \rho$ and diverges if $|T| > \rho$.

Exercises

62. a. Modify the given proof of the Lagrange-Sylvester theorem to obtain the slightly more general version:

   Theorem (Lagrange-Sylvester Mk II):

   If $P^{-1}TP = \overset{m}{\underset{i=1}{\oplus}} J_{n_i}(\lambda_i)$ and $s_{io} = \overset{\infty}{\underset{n=0}{\Sigma}} \alpha_n \lambda^n$ and

   $$s_{ij} = \overset{\infty}{\underset{n=0}{\Sigma}} n(n-1) \cdots (n-j+1)\alpha_n \lambda_i^{n-j} \text{ converge}$$

   for every $0 < j < n_i$ and $1 \le i \le m$, then

   $$\overset{\infty}{\underset{n=0}{\Sigma}} \alpha_n T^n = P\left(\overset{m}{\underset{i=1}{\oplus}} \overset{n_i-1}{\underset{j=0}{\Sigma}} \frac{s_{ij}}{j!} J_{n_i}^j(0)\right) P^{-1}.$$

   b. Show that $\overset{\infty}{\underset{n=1}{\Sigma}} \dfrac{1}{n^2} \begin{bmatrix} 1 & 4 \\ -1 & -3 \end{bmatrix}^n$ converges.

63. State and prove an extension of the Lagrange-Sylvester theorem that begins: Suppose (1) $\rho > 0$ is the radius of convergence of $\Sigma \alpha_n (\tau - \alpha)^n$ and $f(\tau) = \Sigma \alpha_n (\tau - \alpha)^n$ for all $|\tau - \alpha| < \rho$; (2) $|\lambda - \alpha| < \rho$ for all eigenvalues λ of T, ...

14. MATRIX FUNCTIONS AND EXTENSIONS OF SCALAR FUNCTIONS

In scalar analysis you saw functions defined by power series; we can of course do the same thing for matrices. If $\mathcal{T}$ is the

set of all matrices T such that $\sum_{n=0}^{\infty} \alpha_n T^n$ converges, and if
we define f by

$$f(T) = \sum_{n=0}^{\infty} \alpha_n T^n \quad \text{for all } T \ \varepsilon \ \mathcal{C}$$

then f is a matrix-valued function of one matrix variable.
If ρ is the radius of convergence of $\sum \alpha_n \tau^n$, then the
corollary to theorem 3 tells us that f is defined for all
$|T| < \rho$.

If we start with a scalar function (e.g., the arctangent
function) and represent it by a MacLaurin series (e.g.,
$\tan^{-1}(\tau) = \sum_{n=0}^{\infty} (-1)^n/(2n + 1)\tau^{2n+1}$, it is customary to use
the matrix function obtained from the power series thus,
$\tan^{-1}(T) = \sum_{n=0}^{\infty} \frac{(-1)^n}{2n+1} T^{2n+1}$ for all $|T| < 1$. The matrix

function in this sutuation is sometimes called *the matrix
extension of the scalar function*, and we usually refer to it
as the matrix-(function name). (In the example, the matrix
extension of the arctangent function is called the matrix
arctangent.)

One important matrix extension is the matrix exponential:
$e^T = \sum_{n=0}^{\infty} (1/n!)T^n$. This function is defined for all

matrices because $\sum (1/n!)\tau^n$ converges for all τ. If you
prefer to write $\exp(\tau)$ for e^τ then you would write $\exp(T)$
instead of e^T. We shall return to the matrix exponential
in Chap. 3.

Exercises

64. Show that det[exp(A)] = exp[trace(A)] for all A.

65. Prove that exp(A) is nonsingular no matter how A is
    chosen.

66. Prove that $f(A^{tr}) = [f(A)]^{tr}$ when $f(A)$ is the matrix
    extension of $f(\tau)$.

67. Show that $e^{\alpha}B = e^{\alpha I}B$ for all scalars α and all matrices
    B.

68. Show that $e^{I} = eI$.

15. EXTENSION OF SCALAR IDENTITIES

Consider the matrix extensions of sine and cosine

$$\sin T = \sum_{n=0}^{\infty} (-1)^n \frac{1}{(2n+1)!} T^{2n+1} \quad \text{for all } T$$

$$\cos T = \sum_{n=0}^{\infty} (-1)^n \frac{1}{(2n)!} T^{2n} \quad \text{for all } T$$

We know that

$$\sin^2\tau + \cos^2\tau = 1 \quad \text{for all } \tau$$

1. Does $\sin^2 T + \cos^2 T = I$ for all T?
    We know that

    $$\sin 2\tau = 2 \sin \tau \cos \tau \quad \text{for all } \tau$$

2. Does $\sin 2T = 2 \sin T \cos T$ for all T ?

The answer to these questions is "yes." You could try
to prove that by direct computation, but you would encounter
some technical difficulties if you tried to do so because
multiplication of series is rather messy.

There is an indirect method which not only settles these two questions, but it provides a way of proving the general statement about the extension of scalar identities to matrix identities (Theorem 5).

The method is based on the fact that the sum of the matrix extension of two functions is the matrix extension of their sum and the product of two matrix extensions is the matrix extension of their product.

Lemma. If $f(\tau) = \sum_{j=0}^{\infty} \alpha_j \tau^j$, $g(\tau) = \sum_{j=0}^{\infty} \beta_i \tau^j$, $s(\tau) = f(\tau) + g(\tau)$, and $p(\tau) = f(\tau)g(\tau)$ for all $|\tau| < \rho$, then (1) $s(T) = f(T) + g(T)$ and (2) $p(T) = f(T)g(T)$ for all $|T| < \rho$.

Proof: Suppose $|T| < \rho$

1. We have

$$s(\tau) = \sum_{j=0}^{\infty} (\alpha_j + \beta_j)\tau^j \quad \text{for all } |\tau| < \rho$$

therefore

$$s(T) = \sum_{j=0}^{\infty} (\alpha_j + \beta_j)T^j$$

by the Corollary to Theorem 3

$$= \sum_{j=0}^{\infty} \alpha_j T^j + \sum_{j=0}^{\infty} \beta_j T^j \quad \text{[by Exercise 59(a)]}$$

$$= f(T) + g(T).$$

2. By theorem 3 we have

$$f(T)g(T) = P\left(\bigoplus_{i=1}^{m} \sum_{j=0}^{k} \frac{f^{(j)}(\lambda_i)}{j!} U_{r_i}^j \right) P^{-1} P \left(\bigoplus_{i=1}^{m} \sum_{n=0}^{k} \frac{g^{(n)}(\lambda_i)}{n!} U_{r_i}^n \right) P^{-1}$$

$$= P\left(\bigoplus_{i=1}^{m} \sum_{j=0}^{k} \sum_{n=0}^{k} \left[\frac{f^{(j)}(\lambda_i)g^{(n)}(\lambda_i)}{j!\,n!}\right] U_{r_i}^{j+n}\right)P^{-1}$$

Let $\ell = j + n$

$$= P\left(\bigoplus_{i=1}^{m} \sum_{\ell=0}^{k} \sum_{j=0}^{\ell} \left[\frac{f^{(j)}(\lambda_i)g^{(\ell-j)}(\lambda_i)}{j!(\ell - j)!}\right] U_{r_i}^{\ell}\right)P^{-1}$$

because $U_r^{\ell} = 0$ when $\ell \geq r$. Therefore,

$$\text{(9)}$$

$$f(T)g(T) = P\left(\bigoplus_{i=1}^{m} \sum_{\ell=0}^{k} \sum_{j=0}^{\ell} \binom{\ell}{j}\frac{f^{(j)}(\lambda_i)g^{(\ell-j)}(\lambda_i)}{\ell!} U_{r_i}^{\ell}\right)P^{-1}$$

As you may recall from your past experience with power series

$$f(\tau)g(\tau) = \sum_{\ell=0}^{\infty} \sum_{j=0}^{\ell} \alpha_j \beta_{\ell-j} \tau^{\ell} \quad \text{for all } |\tau| < \rho \quad \text{(10)}$$

(See Churchill (1948), pp. 111–112 or Knopp (1952, p. 86).

Since $p(\tau)$ has a power series expansion valid for all $|\tau| < \rho$, theorem 3 applies to $p(\tau)$ also, and we have

$$p(T) = P\left(\bigoplus_{i=1}^{m} \sum_{\ell=0}^{k} \frac{p^{(\ell)}(\lambda_i)}{\ell!} U_{r_i}\right)P^{-1} \quad \text{(11)}$$

According to Leibniz's rule

$$p^{(\ell)}(\tau) = \sum_{j=0}^{\ell} \binom{\ell}{j} f^{(j)}(\tau)g^{(\ell-j)}(\tau)$$

which you can verify by induction on ℓ.

Therefore (2) implies

$$p(T) = P\left(\bigoplus_{i=1}^{m} \sum_{\ell=0}^{k} \left[\sum_{j=0}^{\ell} \binom{\ell}{j}\frac{f^{(j)}(\lambda_i)g^{(\ell-j)}(\lambda_i)}{\ell!}\right] U_{r_i}^{\ell}\right)P^{-1}$$

$$= f(T)g(T) \quad \text{by (1)} \qquad []$$

To illustrate the method, let's return to question 1: Let T be any matrix. There exist α_j and β_j such that $\sin^2 \tau = \sum_{j=0}^{\infty} \alpha_j \tau^j$ and $\cos^2 \tau = \sum_{j=0}^{\infty} \beta_j \tau^j$ for all j. We don't care what the actual values of α_j and β_j are (the point of this method is that we don't have to calculate them, we only need to know they exist). They exist because of Eq. (10) and the fact that $\sin \tau$ and $\cos \tau$ have power series which converge for all τ. The lemma implies that

$$\sin^2(T) = \sum_{j=0}^{\infty} \alpha_j T^j \qquad \cos^2(T) = \sum_{j=0}^{\infty} \beta_j T^j$$

Therefore by Exercise 59(a)

$$\sin^2 T + \cos^2 T = \sum_{j=0}^{\infty} (\alpha_j + \beta_j) T^j \qquad (12a)$$

$$\sin^2 \tau + \cos^2 \tau = \sum_{j=0}^{\infty} (\alpha_j + \beta_j) \tau^j \qquad (12b)$$

But $\sin^2 \tau + \cos^2 \tau = 1$, and hence $\alpha_0 + \beta_0 = 1$, while $\alpha_j + \beta_j = 0$ for all $j > 0$. Consequently, Eq. (12a) implies that $\sin^2 T + \cos^2 T = I$.

As for question 2: There exist δ_j such that $\sin \tau \cos \tau = \sum_{j=0}^{\infty} \delta_j \tau^j$ for all τ; so the lemma implies that

$$\sin T \cos T = \sum_{j=0}^{\infty} \delta_j T^j \qquad (13)$$

But $2 \sin \tau \cos \tau = \sin 2\tau$; so for all τ

$$\sin 2\tau = \sum_{j=0}^{\infty} 2\delta_j T^j$$

and hence

$$\sin 2T = \sum_{j=0}^{\infty} 2\delta_j T^j$$

On the other hand, $2 \sin T \cos T = \sum_{j=0}^{\infty} 2\delta_j T^j$ from Eq. (13).

Therefore $\sin 2T = 2 \sin T \cos T$.

Exercises

69.   Show that $\cos 2T = \cos^2 T - \sin^2 T$ for all T.

70.   Show that $\sin^2 T = \frac{1}{2}(I - \cos 2T)$.

71.   Suppose $f_i(\tau) = \sum_{n=0}^{\infty} \alpha_{in} \tau^n$ for all $|\tau| < \rho$ $(1 \le i \le m)$

    a.   Show that there exist scalars $\gamma_0, \gamma_1, \ldots, \gamma_n, \ldots$

       such that $\prod_{i=1}^{m} f_i(\tau) = \sum_{n=0}^{\infty} \gamma_n \tau^n$ for all $|\tau| < \rho$ (use

       induction on m and Eq. 10).

    b.   Show that $\prod_{i=1}^{m} f_i(T) = \sum_{n=0}^{\infty} \gamma_n T^n$ for all $|T| < \rho$ (use

       induction on m and the preceding lemma). This

       establishes that $\prod_{i=1}^{m} f_i(T)$ is the matrix extension

       of $\prod_{i=1}^{m} f_i(\tau)$.

72.   If $g_1(T), \ldots, g_\ell(T)$ are matrix extensions of

$g_1(\tau), \ldots, g_\ell(\tau)$ (defined for all $|T| < \rho$), show that

$\sum_{i=1}^{\ell} \delta_i g_i(T)$ is the matrix extension of $\sum_{i=1}^{\ell} \delta_i g_i(\tau)$.

    Theorem 5. Suppose $f_1, f_2, \ldots, f_m$ are scalar functions

of one variable, $\Phi(\tau_1, \tau_2, \ldots, \tau_m)$ is a polynomial in m variables, and $0 < \rho \leq \infty$.   If

$$f_i(\tau) = \sum_{n=0}^{\infty} \alpha_{in} \tau^n \quad \text{for all} \quad |\tau| < \rho \quad (1 \leq i \leq m)$$

and

$$\Phi[f_1(\tau), f_2(\tau), \ldots, f_m(\tau)] = 0 \quad \text{for all} \quad |\tau| < \rho$$

then

$$\Phi[f_1(T), f_2(T), \ldots, f_m(T)] = 0 \quad \text{for all} \quad |T| < \rho$$

(*Note:* In our example we had (1) $\Phi(\tau_1, \tau_2) = \tau_1^2 + \tau_2^2 - 1$
and (2) $\Phi(\tau_1, \tau_2, \tau_3) = 2\tau_1\tau_2 - \tau_3$ with $f_1(\tau) = \sin \tau$,
$f_2(\tau) = \cos \tau$ and $f_3(\tau) = \sin 2\tau$.)

   Proof: Let $\Gamma(\tau) = \Phi[f_1(\tau), \ldots, f_m(\tau)]$; then $\Gamma(\tau)$ is a
linear combination of products of powers of the $f_i(\tau)$, and
so, by Exercises 71 and 72, $\Gamma(T) = \Phi[f_1(T), \ldots, f_n(T)]$ is
the matrix extension of $\Gamma(\tau)$.   Therefore, (assuming $|\tau| < \rho$
and $|T| < \rho$) there are $\gamma_0, \gamma_1, \ldots, \gamma_n, \ldots$ such that

$\Gamma(T) = \sum_{n=0}^{\infty} \gamma_n T^n$ and $\Gamma(\tau) = \sum_{n=0}^{\infty} \gamma_n \tau^n$.   On the other hand,

$\Gamma(\tau) = 0$ for all $|\tau| < \rho$; so all $\gamma_n = 0$ (because $\sum_{n=0}^{\infty} 0\tau^n$
is the only power series representation of 0) and hence

$\Gamma(T) = \sum_{n=0}^{\infty} 0T^n = 0$.

   Therefore $\Phi[f_1(T), f_2(T), \ldots, f_n(T)] = 0$ when $|T| < \rho$.

Exercises

73.   Prove that $(I - T) \sum_{n=0}^{\infty} T^n = I$ for all $|T| < 1$ [and hence
$\sum_{n=0}^{\infty} T^n = (I - T)^{-1}$ whenever $|T| < 1$].

74. Use Exercise 73 to compute $\sum\limits_{n=0}^{\infty} \begin{bmatrix} 0.1 & 0.7 \\ 0.3 & 0.6 \end{bmatrix}^n$. (Compare with the method of calculation used at the end of Sec. 5.)

75. Prove that $\sum\limits_{n=0}^{\infty} nT^n = T(I - T)^{-2}$ when $|T| < 1$.

76. Evaluate $\sum\limits_{n=0}^{\infty} \dfrac{n^2}{10^n} \begin{bmatrix} 1 & 2 \\ 8 & 1 \end{bmatrix}^n$.

77. Show that $\sum\limits_{n=0}^{\infty} (A^n - L)$ converges iff $\lim\limits_{n\to\infty} A^n = L$. (*Hint:* see Exercise 53.)

16. EVALUATION OF MATRIX FUNCTIONS

The Lagrange-Sylvester theorem provides a method of evaluation of matrix functions, as in Example 3 of Sec. 13, but you have to know the Jordan form of A and the matrix P that puts A into Jordan form to use the theorem directly. This is a strong limitation on the effectiveness of a direct application of the theorem as a method of evaluation.

Example 1 (Direct Application): $A = \begin{bmatrix} 1 & 4 \\ 3 & 2 \end{bmatrix} = P \begin{bmatrix} 5 & 0 \\ 0 & -2 \end{bmatrix} P^{-1}$, where $P = \begin{bmatrix} 1 & -4 \\ 1 & 3 \end{bmatrix}$. Lagrange-Sylvester implies

$$e^A = P \begin{bmatrix} e^5 & 0 \\ 0 & e^{-2} \end{bmatrix} P^{-1} = \frac{1}{7} \begin{bmatrix} 3e^5 + 4e^{-2} & 4e^5 - 4e^{-2} \\ 3e^5 - 3e^{-2} & 4e^5 + 3e^{-2} \end{bmatrix}$$

At the end of Sec. 7 we discussed a method of applying the Cayley-Hamilton theorem to evaluate matrix polynomials. That method can be extended by means of the Lagrange-Sylvester theorem: Let $\ell(\lambda)$ denote the multiplicity of λ as a root

of the minimal† polynomial of A, that is, $m_A(\tau) =$
$\prod_{i=1}^{q} (\tau - \lambda_i)^{\ell(\lambda_i)}$. Suppose we can find a polynomial $r(\tau)$ of

degree less than the degree of the minimal polynomial

satisfying

$$r^{(j)}(\lambda) = f^{(j)}(\lambda) \tag{14}$$

. for each eigenvalue λ of A and each $0 \le j < \ell(\lambda)$. The

Lagrange-Sylvester theorem would then imply that $r(A) = f(A)$,

because $\ell(\lambda_i) \le n_i$ for i = 1, 2, ..., m.

Example 2. Evaluate e^A if A = $\begin{bmatrix} 1 & 4 \\ 3 & 2 \end{bmatrix}$:

$m_A(\tau) = c_A(\tau) = (\tau - 5)(\tau + 2)$, so we try to find $r(\tau) =$
$\alpha + \beta\tau$ satisfying

$$r(5) = e^5$$
$$r'(-2) = e^{-?}$$

that is,

$$\alpha + \beta 5 = e^5$$
$$\alpha - \beta 2 = e^{-2}$$

We obtain

$$r(\tau) = \frac{1}{7}(2e^5 + 5e^{-2}) + \frac{1}{7}(e^5 - e^{-2})\tau$$

---

† If it's more convenient, you can try to solve Eq. (14) for
each $0 \le j < p(\lambda)$, where $p(\lambda)$ is the multiplicity of λ as
a root of the characteristic polynomial, but the minimal
polynomial is easier to find when k is large and $\ell(\lambda)$ can
be very much smaller than $p(\lambda)$, so it is usually better to
use the minimal polynomial when k is (even modestly) large.
(See Sec. 17 for a review of minimal polynomials.)

but $e^A = r(A)$; so

$$e^A = \frac{1}{7}(2e^5 + 5e^{-2})\begin{bmatrix} 1 & 0 \\ 0 & 1 \end{bmatrix} + \frac{1}{7}(e^5 - e^2)\begin{bmatrix} 1 & 4 \\ 3 & 2 \end{bmatrix}$$

Example 3.   If $A = \begin{bmatrix} 0 & -1 \\ 4 & 4 \end{bmatrix}$, evaluate $\sin^{-1}(\frac{1}{4})A$:

$c_A(\tau) = (\tau - 2)^2 = m_A(\tau)$, so we try to find $r(\tau) = \alpha + \beta\tau$ satisfying

$$r(2) = \sin^{-1}\frac{1}{2}$$
$$r^{(1)}(2) = \frac{1}{4}(1 - (\frac{2}{4})^2)^{-1/2} \qquad\qquad (15)$$

that is,

$$\alpha + 2\beta = \frac{\pi}{6}$$
$$\beta = \frac{\sqrt{3}}{6}$$

We obtain

$$r(\tau) = \frac{1}{6}[(\pi - 2\sqrt{3}) + \tau\sqrt{3}]$$

but $\sin^{-1}(\frac{1}{4})A = r(A)$; so

$$\sin^{-1}(\frac{1}{4})A = (\frac{1}{6})\begin{bmatrix} \pi - 2\sqrt{3} & -\sqrt{3} \\ 4\sqrt{3} & \pi + 2\sqrt{3} \end{bmatrix}$$

Example 4.   Evaluate $\cos \pi A$ if $A = \begin{bmatrix} -2 & 2 & -2 & 4 \\ -1 & 2 & -1 & 1 \\ 0 & 0 & 1 & 0 \\ -2 & 1 & -1 & 4 \end{bmatrix}$.

In this case $m_A(\tau) = \tau^3 - 4\tau^2 + 5\tau - 2$

$$= (\tau - 1)^2(\tau - 2) \quad (\text{see Sec. 17})$$

so we try to find $r(\tau) = \alpha + \beta\tau + \gamma\tau^2$ satisfying

$$r(2) = \cos 2\pi$$
$$r(1) = \cos \pi \qquad\qquad (16)$$
$$r^{(1)}(1) = -\pi \sin \pi$$

that is,

$$\alpha + \beta 2 + \gamma 4 = 1$$
$$\alpha + \beta + \gamma = -1$$
$$\beta + 2\gamma = 0$$

We obtain

$$r(\tau) = 1 - 4\tau + 2\tau^2$$

but $\cos \pi A = r(A)$; so

$$\cos \pi A = I - 4A + 2A^2 = \begin{bmatrix} -3 & 0 & 0 & 4 \\ 0 & -1 & 0 & 0 \\ 0 & 0 & -1 & 0 \\ -2 & 0 & 0 & 3 \end{bmatrix}$$

Exercises

78. Evaluate e^A if $A = \begin{bmatrix} 0 & -1 \\ 4 & 4 \end{bmatrix}$.

79. Evaluate $\cos \pi \begin{bmatrix} 0 & 0 & 0 & 2 \\ 0 & 1 & 0 & 0 \\ 0 & 0 & 1 & 0 \\ -1 & 0 & 0 & 3 \end{bmatrix}$.

80. Evaluate $\tan^{-1} \frac{1}{4} \begin{bmatrix} -3 & 2 & 1 \\ 1 & -2 & 1 \\ 1 & 2 & -3 \end{bmatrix}$.

81. Suppose $A = \begin{bmatrix} -1 & 2 & -2 & 2 \\ -2 & 2 & -1 & 3 \\ -1 & 0 & 1 & 2 \\ -1 & 1 & -1 & 2 \end{bmatrix}$. Evaluate $\tan^{-1}(A^{30} - \frac{1}{2} A^{20})$.

Will the method we've just presented always work? That is, can we always find a polynomial $r(\tau)$ satisfying Eq. 14?

Fortunately, the answer is yes. To see why, let V be the subspace of $\underline{C}^{k^2}$ spanned by I, A, $\ldots$, A^{k-1} and suppose $\sum\limits_{n=0}^{\infty} \alpha_n A^n$ converges to $f(A)$. The Cayley-Hamilton theorem implies that the partial sums, $\sum\limits_{n=0}^{p} \alpha_n A^n$ all lie in V, therefore their limit, $f(A)$, must also lie in V, and hence $f(A) = \beta_0 I + \beta_1 A + \cdots + \beta_{k-1} A^{k-1}$ for some scalars $\beta_i \, \varepsilon \, \underline{C}$. If we define $r(\tau) = \sum\limits_{i=0}^{k-1} \beta_i \tau^i$, then according to the Lagrange-Sylvester theorem we have

$$r(A) = P\left(\bigoplus_{i=1}^{m} \sum_{j=0}^{n_i-1} \frac{r^{(j)}(\lambda_i)}{j!} U_{n_i}^j \right) P^{-1}$$

and

$$f(A) = P\left(\bigoplus_{i=1}^{m} \frac{f^{(j)}(\lambda_i)}{j!} U_{n_i}^j \right) P^{-1}$$

Comparing entries in $P^{-1}r(A)P$ with those in $P^{-1}f(A)P$, we have $r^{(j)}(\lambda_i) = f^{(j)}(\lambda_i)$ for each $1 \leq i \leq m$ and each $0 \leq j < n_i$. Now $\ell(\lambda) = \max\{n_i : \lambda = \lambda_i\}$, so

$$r^{(j)}(\lambda) = f^{(j)}(\lambda) \tag{17}$$

for each eigenvalue λ of A and each $0 \leq j < \ell(\lambda)$ for this polynomial $r(\tau)$. We've already seen that any solution r to Eq. (17) has $r(A) = f(A)$, so this method will always work.

17. THE MINIMAL POLYNOMIAL REVIEWED

The polynomial $p(\tau) = \sum\limits_{i=0}^{m} \beta_i \tau^i$ is a *minimal polynomial* of the $k \times k$ matrix A iff

1. $p(A) = 0$

2. If $s(\tau)$ is a nonzero polynomial of degree m' and
    $s(A) = 0$, then $m' \geq m$

3. $\beta_m = 1$

In other words, $p(\tau)$ is a monic polynomial of least degree
which has A for a root.

How do we know there is a minimal polynomial for A?
How do we find one? How many are there?

Every Matrix Has a Minimal Polynomial

The finite sequence of matrices $I, A, A^2, \ldots, A^{k^2}$
is linearly dependent (as a sequence of $k^2 + 1$ members of
the k^2-dimensional space of all k x k matrices). Let m be
the smallest index t such that $I, A, \ldots, A^t$ is linearly
dependent. Therefore there exist γ_i, not all zero, such that
$\sum_{i=0}^{m} \gamma_i A^i = 0$. In particular, $\gamma_m \neq 0$ because of the minimality
of m. Therefore

$$A^m + \sum_{i=0}^{m-1} \frac{\gamma_i}{\gamma_m} A_i = 0$$

If we define the polynomial p by

$$p(\tau) = \tau^m + \sum_{i=0}^{m-1} \frac{\gamma_i}{\gamma_m} \tau^i \quad \text{for all } \tau$$

then (1) $p(A) = 0$, (2) no polynomial of degree less than p's
has A for a root, and (3) p is monic; that is, p is a minimal
polynomial for A.

How to Find a Minimal Polynomial for A

The proof of existence of a minimal polynomial suggests
an algorithm for finding one:

1. Try to solve $A = \lambda_0 I$ for λ_0. If there is no solution,

2. Try to solve $A^2 = \lambda_0 I + \lambda_1 A$ for λ_0, λ_1. If there
    is no solution,
$\vdots$

j. Try to solve $A^j = \sum_{i=0}^{j-1} \lambda_i A^i$ for λ_0, ..., λ_{j-1}. If
    there is no solution,
$\vdots$

    (This procedure must stop at some step as seen in
    Sec. 17.A)
$\vdots$

m. $A^m = \sum_{i=0}^{m-1} \lambda_i A^i$;

so

$$p(\tau) = \tau^m - \sum_{i=0}^{m-1} \lambda_i \tau^i$$

Example: Suppose A is as in Example 4 of Sec. 16.

1. Try to solve $A = \lambda I$ for λ. There is no solution
    (otherwise, e.g., $-2 = \lambda$ and $2 = \lambda$).

2. Try to solve $A^2 = \lambda_0 I + \lambda_1 A$. If there is a solution
    then comparing the first rows of both sides:

    $[-6, 4, -4, 10] = [\lambda_0 - 2\lambda_1, 2\lambda_1, -2\lambda_1, 4\lambda_1]$
    and we'd have $4 = 2\lambda_1$ and $10 = 4\lambda_1$; so there
    is no solution.

3. Try to solve $A^3 = \lambda_0 I + \lambda_1 A + \lambda_2 A^2$. If there is a
    solution, then comparing first rows of both sides:
    $[-12, 6, -6, 20] = [\lambda_0 - 2\lambda_1 - 6\lambda_2, 2\lambda_1 + 4\lambda_2,$
    $4\lambda_1 + 10\lambda_2]$
    and we'd have $\lambda_2 = 4$, $\lambda_1 = -5$, and $\lambda_0 = 2$. Is

there a solution? Yes, because $2I - 5A + 4A^2 = A^3$. Therefore $\tau^3 - 4\tau^2 + 5\tau - 2$ is a minimal polynomial for A.

Every Matrix Has at Most One Minimal Polynomial

Suppose p' and p are minimal polynomials for A. Dividing p' by p, we find polynomials q (the quotient polynomial) and r (the remainder polynomial) such that

$$p'(\tau) = p(\tau)q(\tau) + r(\tau) \quad \text{for all } \tau$$

Therefore, $r(A) = p'(A) - p(A)q(A) = 0$ by part (1) of the definition of minimal polynomial. Therefore part (2) of that definition ensures that r is the zero polynomial because the remainder's degree is always smaller than the divisor's. Consequently, $p'(\tau) = p(\tau)q(\tau)$. Part (2) implies that p' and p have the same degree, therefore $q(\tau)$ is a constant polynomial $[q(\tau) = \alpha$ for all $\tau]$. Part (3) implies that $\alpha = 1$; so $p'(\tau) = p(\tau)$ for all τ, and hence every matrix has at most one minimal polynomial.

Exercises

82. Show that the minimal polynomial of A is a divisor of every polynomial having A for a root, i.e., if m_A is the minimal polynomial of A and f is any polynomial such that $f(A) = 0$ then there is some polynomial q such that

$$f(\tau) = m_A(\tau)q(\tau) \quad \text{for all } \tau.$$

83. Show that λ is an eigenvalue of A iff $m_A(\lambda) = 0$.

84. Find the minimal polynomial of

    a. I   b. 0

    c. $J_n(\lambda)$ d. $\displaystyle\bigoplus_{i=1}^{m} J_{n_i}(\lambda_i)$.

CHAPTER 2

NONNEGATIVE MATRICES

1. INTRODUCTION AND DEFINITIONS

   Example 1:   An association of four regions R_1, R_2, R_3, R_4 trade in a certain nonrenewable commodity. Assume that the commodity is shipped between and within the regions according to the following matrix of fractions:

$$C = \begin{bmatrix} 0.1 & 0.3 & 0 & 0 \\ 0.7 & 0.6 & 0 & 0 \\ 0.1 & 0 & 1 & 0 \\ 0.1 & 0.1 & 0 & 1 \end{bmatrix}$$

where c_{ij} is the fraction of the commodity present at R_j which is shipped each day to R_i. Suppose we know the amounts v_1, v_2, v_3, v_4 of the commodity present in R_1, R_2, R_3, R_4 respectively today. What will the distribution of the commodity be like two weeks or a month from now?

   Since $c_{ij}v_j$ is the amount shipped today from R_j to R_i, $\sum_{j=1}^{4} c_{ij}v_j$ is the amount present at R_i tomorrow. In general, $C^n v$ gives the distribution of the commodity in each region n days from today.

$$C = \begin{bmatrix} X & 0 \\ Y & I \end{bmatrix}, \text{ where } X = \begin{bmatrix} 0.1 & 0.3 \\ 0.7 & 0.6 \end{bmatrix}, \; Y = \begin{bmatrix} 0.1 & 0 \\ 0.1 & 0.1 \end{bmatrix},$$

$0 = \begin{bmatrix} 0 & 0 \\ 0 & 0 \end{bmatrix}$, and $I = \begin{bmatrix} 1 & 0 \\ 0 & 1 \end{bmatrix}$. Therefore [see Lipshutz (1968) pp. 45-46.]

$$C^n = \begin{bmatrix} X^n & 0 \\ Y \sum_{j=0}^{n-1} X^j & I \end{bmatrix} \quad \text{for all } n$$

Since $|X| < 1$, it follows that $I - X$ is invertible and

$$(I - X)^{-1} = \sum_{n=0}^{\infty} X^n$$

This is the result of Exercise 73 of Chap. 1, but we can also prove it directly:

$$\sum_{n=0}^{m-1} X^n (I - X) = I - X^m$$

$I - X$ is invertible, because 1 cannot be an eigenvalue of X. Therefore

$$\sum_{n=0}^{m-1} X^n = (I - X)^{-1} - X^m (I - X)^{-1}$$

But

$$\lim_{m \to \infty} X^m = 0 \quad \text{(Exercise 51 of Chap. 1)}$$

so

$$\sum_{n=0}^{\infty} X^n = (I - X)^{-1}$$

Therefore

$$\sum_{n=0}^{\infty} X^n = (I - X)^{-1} = \begin{bmatrix} 0.9 & -0.3 \\ -0.7 & 0.4 \end{bmatrix}^{-1} = \begin{bmatrix} \frac{8}{3} & 2 \\ \frac{14}{3} & 6 \end{bmatrix}$$

Notice how much easier this caluclation is than the one used

in Sec. 5 to compute $\sum_{n=0}^{\infty} \begin{bmatrix} 0.1 & 0.7 \\ 0.3 & 0.6 \end{bmatrix}^n$. We didn't even need to

find the eigenvalues of X, since $|X| \leq \max_{1 \leq j \leq k} \sum_{i=1}^{k} |x_{ij}|$ for

all X -- can you prove this? -- so, in our case, $\left| \begin{bmatrix} 0.1 & 0.3 \\ 0.7 & 0.6 \end{bmatrix} \right| \leq$

0.9 < 1. Consequently,

$$\lim_{n \to \infty} C^n = \begin{bmatrix} 0 & 0 & 0 & 0 \\ 0 & 0 & 0 & 0 \\ \frac{4}{15} & \frac{1}{5} & 1 & 0 \\ \frac{11}{15} & \frac{4}{5} & 0 & 1 \end{bmatrix}$$

$$\lim_{n \to \infty} C^n v = \left[0, \ 0, \ \frac{4v_1 + 3v_2 + 15v_3}{15}, \ \frac{11v_1 + 12v_2 + 15v_4}{15} \right]^{tr}$$

So eventually R_3 and R_4 will have all the commodity with R_4
getting the lion's share. This happens fairly quickly, in
fact C^n approximates its limit to one decimal place after 19
days have elapsed.

   Example 2 (The Subsidy Problem): Certain industries,
each manufacturing a different product, use each others'
products as well as their own for production. The government
is willing to grant them a joint subsidy of a certain amount.
The industries would like the money to be distributed
equitably, that is, none of them wants to spend more of its

share of the subsidy for its own production costs than it
receives of the subsidy money for its own product. The
government, on its part, wants them to produce as much as
possible. The industries will abide by production quotas if
they will lead to an equitable distribution of costs.

Given that the cost to the jth industry for using the
i^{th} industry's product is c_{ij} dollars per unit of its own
product (naturally we assume that all $c_{ij} > 0$) and that
there are k industries all together, can the subsidy be
distributed equitably? To what extent does the possibility
of doing so depend on the relative costs c_{ij}?

Let's examine the problem more closely: Suppose the
i^{th} industry produces u_i units yearly. The jth industry
pays out a total of $\sum_i c_{ij} u_j$ subsidy dollars yearly and, on

the other hand, it receives $\sum_i c_{ji} u_i$ subsidy dollars yearly
in payment from the industries; so its net annual subsidized
production cost (v_j) is given by

$$v_j = \gamma_j u_j - \sum_i c_{ji} u_i$$

where $\gamma_j = \sum_i c_{ij}$. To be equitable, we must have no $v_j > 0$,

but this implies that each $v_j = 0$, and we have reduced the
subsidy problem to finding a system of production quotas
$u_1, u_2, \ldots, u_k$ so that

Condition (1): $u_j = \sum_{i=1}^{k} \dfrac{c_{ji}}{\gamma_j} u_i$ $(j = 1, 2, \ldots, k)$

subject to Restriction (1), $u_j > 0$ $(j = 1, 2, \ldots, k)$ and

Restriction (2), $\sum\limits_j u_j \geq \sum\limits_j x_j$ for any x such that $0 < x_j =$

$= \sum\limits_{i=1}^{k} \dfrac{c_{ji}}{\gamma_j} x_i$ for $j = 1, 2, \ldots,$ and k.

   Condition (1) ensures that the system is equitable.

   Restriction (1) ensures that the system is feasable.

   Restriction (2) ensures that the system is optimal.

   Must Condition (1) have solutions? If so, will any
of them satisfy restriction (1)? As you know, condition (1)
can be written as a homogeneous system of k linear equations
in k unknowns. We can therefore expect nonzero solutions to
condition (1) if we were sure the rank of the corresponding
homogeneous system were less than k. But even if there were
nonzero solutions to condition (1) it wouldn't appear likely
that there would have to be one satisfying restriction (1).
It would seem that the issue depends on the costs c_{ij}: that
there should be solutions to condition (1) for some C but not
for others. It would be natural to expect that even if
condition (1) had solutions, none might satisfy restriction
(1).

   In 1905, the German mathematician, O. Perron, proved a
basic theorem about real matrices A all of whose entries are
positive. Among many other things his theorem guarantees
that for such matrices A, the equation $Ax = |A|x$ has a
solution x with each $x_i > 0$. So if we put $a_{ji} = \dfrac{c_{ji}}{\gamma_j}$, then

$\sum\limits_{i=1}^{k} \dfrac{c_{ji}}{\gamma_j} x_i = |A|x_j$ for each $j = 1, 2, \ldots, k.$ On the other hand, letting $g = [\gamma_1, \gamma_2, ..., \gamma_k]^{tr}$ we have $g^{tr}A = g^{tr}$; therefore, $g^{tr}x = g^{tr}Ax = g^{tr}|A|x = |A|g^{tr}x$. Since the scalar $g^{tr}x$ does not equal zero, we must have $|A| = 1$.

Any vector of the form τx where $\tau > 0$, is a solution to condition (1) and satisfies restriction (1). Perron's Theorem also implies that the space of all solutions y to $Ay = |A|y$ is one-dimensional. Therefore, the only solutions to condition (1) subject to restriction (1) are positive multiples of x. $\sum_{i=1}^{k} \tau x_i$ is the total production under the quota system given by τx. The total cost of production under this system is $\sum_{j} \gamma_j(\tau x_j)$, but this cannot exceed σ, the government subsidy, so we choose the quota system

$$u_i = \frac{\sigma}{\sum_j \gamma_j x_j} x_i \quad (i = 1, 2, ..., k)$$

and obtain a solution to condition (1) satisfying restriction (1) which is also optimal. In this system the ith industry receives $\gamma_i u_i$ as its share of the total subsidy.

So Perron's theorem enabled us to answer the questions raised before: the subsidy problem can *always* be solved *no matter what* the costs c_{ij} are. The theorem also shows that there is only one solution to the problem.

You have just seen two examples involving nonnegative matrices, that is of real matrices with nonnegative entries. We tried to suggest how these matrices turn up in applications. These examples are by no means isolated. Nonnegative matrices occur frequently enough in the description of problems, especially in probability, statistics and economics to warrant special attention.

If M and N are k × ℓ real matrices, we write M ≥ N iff $m_{ij} \geq n_{ij}$ (for all i, j) and M > N iff $m_{ij} > n_{ij}$ (for all i, j).

If M ≥ 0, we say that M is *nonnegative*.

If M > 0, we say that M is *positive*.

Thus $\begin{bmatrix} 1 & 2 \\ 3 & 4 \end{bmatrix}$ and $\begin{bmatrix} 1 \\ 2 \\ 3 \end{bmatrix}$ are positive, while $\begin{bmatrix} 1 & 0 \\ 3 & 4 \end{bmatrix}$ and $\begin{bmatrix} 1 \\ 0 \\ 3 \end{bmatrix}$ are nonnegative (but not positive). So you see that although a nonnegative real number which isn't positive must be zero, the same isn't true for vectors and matrices.

Exercises

1. If $0 \neq x \geq 0$ and A > 0, show that Ax > 0.

2. Find an example of a nonzero, nonnegative, 2 × 2 matrix A and a nonzero, nonnegative, 2 × 1 vector x such that Ax = 0.

3. If A > 0 and z ≥ w, show that Az ≥ Aw, with equality iff z = w.

A sequence of theorems will be presented in Sec. 2 which will provide a proof of Perron's theorem. Some of these results are useful in themselves and all of them are either parts of Perron's theorem or fairly immediate consequences of it. We have chosen this anachronistic way of presenting Perron's theorem (many methods are now available since the theorem first appeared) because it is the briefest matrix-theoretic way we have found, and because it should suggest some of the ways in which the theorem can be used.

2. THE BASIC THEORY OF POSITIVE MATRICES

Throughout this section the matrix A is a k × k *positive* matrix.

   Wielandt's Lemma:

   1. The spectral radius of A is a positive eigenvalue
       of A having a positive eigenvector.
   2. The modulus of any other eigenvalue of A is smaller
       than the spectral radius.

   Proof: Suppose μ is an eigenvalue of maximal modulus, then $|\mu| = |A|$ and $Ax = \mu x$ for some complex nonzero vector x.

   If we define $p = [\,|x_1|,\ |x_2|,\ \ldots,\ |x_k|\,]^{tr}$, then (we'll now show that) p is the required positive eigenvector. For each $1 \le i \le k$, we have $\mu x_i = \sum_j a_{ij}x_j$; therefore,

$$|A|p_i = |\mu x_i| \le \sum_j a_{ij}|x_j| = \sum_j a_{ij}p_j$$

and hence $|A|p \le Ap$. Consequently, $(A - |A|I)p \ge 0$. We have to show that equality holds.

   We argue by contradiction: Suppose z denotes $(A - |A|I)p$ and $z \ne 0$. According to Exercise 1, $Az > 0$. Since $Ap > 0$ also, it follows that we can find a positive number ε so small that $Az \ge \varepsilon Ap$. Now $A(A - |A|I)p = Az$

$$A^2 p = Az + |A|Ap$$
$$\geq \varepsilon Ap + |A|Ap$$
$$\geq (\varepsilon + |A|)Ap$$

Let $B = [(\varepsilon + |A|)^{-1}A]$, then $BAp \geq Ap$. Applying Exercise 3 $n - 1$ times we obtain $B^n Ap \geq Ap$ for all $n \geq 1$. Now $|B| = |(\varepsilon + |A|)^{-1}||A|$, because for any scalar τ and matrix X we have $|\tau X| = |\tau||X|$. But $\varepsilon > 0$; so $|B| < 1$, and hence (Chap. 1, Exercise 51), $\lim_{n \to \infty} B^n = 0$. Therefore, $0 \geq Ap$ but $Ap > 0$. This contradiction establishes the equality, $Ap = |A|p$. We noted before that $Ap > 0$, so both $|A|$ and p must be positive. This proves (1).

In order to see why (2) is so, suppose λ is an eigenvalue of maximal modulus with corresponding eigenvector y. By applying the same argument used above for μ and x to λ and y and letting $q_i = |y_i|$, we have both

$$|A|q_1 = \left|\sum_j a_{1j}y_j\right| \leq \sum_j a_{1j}|y_j| = \sum_j a_{1j}q_j = (Aq)_1$$

and

$$|A|q = Aq$$

Therefore $\left|\sum_j a_{1j}y_j\right| = \sum_j a_{1j}|y_j|$, but the numbers a_{11}, a_{12}, ..., a_{1k} are all positive real numbers so the arguments of all the y_j must be the same, i.e., in writing y_j in polar form, there is real θ independent of j such that $y_j = |y_j|e^{i\theta}$ for all j. Therefore, $y = e^{i\theta}q$. Thus, y is a nonzero scalar multiple of an eigenvector corresponding to the

eigenvalue $|A|$ so v is also an eigenvector for $|A|$. Since
y is an eigenvector for λ as well, it must be that $\lambda = |A|$.

Exercise

4. Find a matrix M such that $|M|$ isn't an eigenvalue of M.

   Theorem 1. $|A|^{-1}A$ is power convergent.

   Proof: Let $B = |A|^{-1}A$ then $|B| = 1$. Applying Wielandt's
Lemma to B, we see that 1 is an eigenvalue of B and $|\lambda| < 1$
for all eigenvalues $\lambda \neq 1$. We'll now show that B is *power
bounded* (see Chap. 1, Exercises 54-56).

   Let y be a positive eigenvector for the eigenvalue 1,
then $B^n y = y$ for all n, and hence, letting $y_s = \max_i y_i$ and
$y_t = \min_i y_i$, we have

$$y_s \geq y_i = \sum_\ell b_{i\ell}^{(n)} y_\ell \geq b_{ij}^{(n)} y_j \geq b_{ij}^{(n)} y_t \quad \text{for all } i, j$$

Therefore, $b_{ij}^{(n)} < \dfrac{y_s}{y_t}$ for all n. Since $b_{ij}^{(n)} > 0$ for all n,

we see that B is power bounded. You should have shown (Chap.
1, Exercise 56) that this implies that the Jordan blocks in
the Jordan form for B corresponding to the eigenvalues of
modulus 1 are all 1×1. Since $|\lambda| < 1$ for eigenvalues
$\lambda \neq 1$, we see (Chap. 1, theorem 2) that B is power convergent.

   Corollary. $|A|$ is a simple[†] eigenvalue of A.

[†] $|A|$ is a root of multiplicity *one* of $c_A(\tau)$.

Proof: In the proof of the previous theorem you saw that the Jordan form for $B = |A|^{-1}A$ is $J = I_n \oplus J_{n_1}(\lambda_1) \oplus \ldots \oplus J_{n_\ell}(\lambda_\ell)$ and $|\lambda_i| < 1$ for $i = 1, 2, \ldots, \ell$. The multiplicity of A as an eigenvalue of A = (the multiplicity of 1 as an eigenvalue of B) = n. So we have to show that n = 1; We have

$$J - I = 0_n \oplus J_{n_1}(\lambda_1 - 1) \oplus J_{n_2}(\lambda_2 - 1) \oplus \cdots \oplus J_{n_\ell}(\lambda_\ell - 1)$$

Therefore

$$n = \text{nullity}(J - I) \quad (\text{as no } \lambda_i = 1)$$
$$= \text{nullity}(B - I) \quad (\text{Because } J \sim B)$$

We also have $Bp = p$ for some $p > 0$ so $n \geq 1$. If $n > 1$, then $Bq = q$ for some real vector q which isn't a scalar multiple of p: Let

$$\tau = \max_i \frac{q_i}{p_i}$$

then choose j so that $\frac{q_j}{p_j} = \tau$. Now $\tau p \geq q$. Exercise 1 implies $B(\tau p - q) > 0$, because $\tau p - q \neq 0$. Therefore,

$$\tau Bp - Bq > 0$$

thus

$$\tau p - q > 0$$

so

$$\tau p > q$$

and hence

$$\tau p_j > q_j$$

which contradicts the choice of j. Therefore, n = 1. []

Exercise

5. The set of all vectors x satisfying Mx = λx is a subspace
    since it is the nullspace of M - λI. If λ is an eigenvalue
    of M, then this space is called the *eigenspace* of λ.

    a. If λ is a simple eigenvalue, show that its eigenspace
        is one-dimensional.

    b. Show that the converse of (a) isn't necessarily true.

Perron's Theorem. If A > 0, then |A| is a simple
eigenvalue of A, |A| has a positive eigenvector, and
|A| > |λ| for all eigenvalues λ of A other than |A|.

Proof: This is a restatement of Wielandt's lemma and
the corollary to theorem 1. []

3. POWERS OF POSITIVE MATRICES

Theorem 2. If A > 0, p is any positive eigenvector of
A corresponding to |A|, and q is any positive eigenvector of
A^{tr} corresponding to |A| (= $|A^{tr}|$), then

$$\lim_{n \to \infty} (|A|^{-1}A)^n = (q^{tr}p)^{-1}pq^{tr}$$

Proof: Let B = $|A|^{-1}A$. Theorem 1 implies that $\lim_{n \to \infty} B^n$
exists, call it L. If $L^{(j)}$ is the jth column of L, then
$L^{(j)} = BL^{(j)}$, because L = $\lim_{n \to \infty} B^{n+1}$ = B $\lim_{n \to \infty} B^n$ = BL.

Therefore the columns of B are all in the eigenspace of the simple eigenvalue 1 of B. This eigenspace is one-dimensional [Exercise 5(a)], and p is in it; therefore, $L^{(j)}$ must be a scalar multiple of p, that is, $L^{(j)} = r_j p$ for some scalar r_j, and hence

$$L = \begin{bmatrix} r_1 p_1 & r_2 p_1 & \cdots & r_k p_1 \\ r_1 p_2 & r_2 p_2 & \cdots & r_k p_2 \\ \vdots & \vdots & & \vdots \\ r_1 p_k & r_2 p_k & \cdots & r_k p_k \end{bmatrix}$$

If we put $r = [r_1, r_2, \ldots, r_k]^{tr}$, then we have $L = pr^{tr}$. But $q^{tr}L = q^{tr}$, because $(B^{tr})^n q = q$ for all $n \geq 1$; so $q^{tr} pr^{tr} = q^{tr}$; therefore, $q^{tr} p \neq 0$ and $r^{tr} = (q^{tr} p)^{-1} q^{tr}$, and hence $L = (q^{tr} p)^{-1} pq^{tr}$. []

Exercises

6. Let $A = \begin{bmatrix} 7 & 2 & 2 \\ 2 & 1 & 1 \\ 4 & 2 & 2 \end{bmatrix}$.

    a. Find $|A|$.

    b. Find $\lim_{n \to \infty} (|A|^{-1} A)^n$ using theorem 2.

    c. Compute $A^8 \begin{bmatrix} 1 \\ 1 \\ 1 \end{bmatrix}$. Take $v = A^8 \begin{bmatrix} 1 \\ 1 \\ 1 \end{bmatrix}$ as an approximation

        to an eigenvector for A. Estimate $|A|$ by taking the
average of the values $\dfrac{(Av)_i}{v_i}$ $(i = 1, 2, 3)$.

7. Let $A = \begin{bmatrix} 1 & 2 & 1 \\ 2 & 2 & 1 \\ 3 & 2 & 1 \end{bmatrix}$. Approximate $|A|$ by the method
suggested in 6(c). You might use a computer to do this.

8.   Suppose M is a k × k matrix, not necessarily nonnegative, $\lambda > 0$ is a simple eigenvalue of M, and $\lambda > |\mu|$ for all other eigenvalues μ of M.

   a.   Show that $\lim_{n \to \infty} (\lambda^{-1}M)^n$ exists.

   b.   Show that $\lim_{n \to \infty} (\lambda^{-1}M)^n = (x^{tr}y)^{-1}yx^{tr}$, where y is any

        eigenvector of M for λ and x is any eigenvector of M^{tr} for λ.

9.   a.   Given that $M = \begin{bmatrix} 1 & -3 & 3 \\ 3 & -5 & 3 \\ 6 & -6 & 4 \end{bmatrix}$ has $(\tau + 2)^2(\tau - 4)$ as its

        characteristic polynomial, find $\lim_{n \to \infty} (\frac{1}{4}M)^n$ by using Exercise 8(b).

   b.   Use a computer to see how good $(\frac{1}{4}M)^{16}$ is as an approximation to the limit found in (a).

4.   AN APPLICATION

Let's return to the first example, given in Sec. 1 of this chapter, of an association trading in a certain nonrenewable commodity.

   Assume that α units of the commodity are shipped between and within the regions according to the matrix

$$C = \begin{bmatrix} 0.1 & 0.2 & 0.1 & 0.1 \\ 0.7 & 0.6 & 0.1 & 0.1 \\ 0.1 & 0.1 & 0.7 & 0.1 \\ 0.1 & 0.1 & 0.1 & 0.7 \end{bmatrix}$$

where c_{ij} is the fraction of the commodity present in R_j which is shipped each day to R_i. In addition, suppose that the α units of the commodity are distributed so that there are v_1, v_2, v_3, v_4 units in regions R_1, R_2, R_3, R_4, respectively,

today. How will the commodity be distributed two weeks or a
month from now? We saw before that the vector $C^n v$ gives us
the amounts present in each region at the end of n days.

Let's apply theorem 2 to see what happens to $C^n v$ when
n is large: If $q = [1, 1, 1, 1]^{tr}$, then $C^{tr}q = q$ so $|C| \geq 1$.
Exercise 10 states that $|C| \leq \max_j (\sum_{i=1}^{4} c_{ij}) = 1$ so $|C| = 1$.
As you may verify: $Cp = p$, when $p = [6, 16, 11, 11]^{tr}$ (p was
found by solving $(10C - I)x = 0$). Applying our theorem we
obtain:

$$\lim_{n\to\infty} C^n v = (q^{tr}p)^{-1}pq^{tr}v$$

$$= \frac{\alpha}{44} [6, 16, 11, 11]^{tr}$$

Notice that this limit depends only on C and α and not at
all on v.

This tells us that regardless of how the α units were
distributed among the four regions initially, in the long
run region R_i will have $(\alpha p_i)/44$ of them (i = 1, 2, 3, and
4).

If the third region had none of the commodity initially,
or even if it had all of it initially, in the long run the
third region would have $\alpha/4$ of them. This situation contrasts
strongly with that of the original example, where in the long
run R_3 would have $4v_1 + 3v_2/15$ units if it had none of them
initially, but it would always have all of them if it began
with all α of them.

The reason for this strong contrast is in the fact that
the new matrix C is positive. We would have reached the

same conclusion (that $\lim\limits_{n \to \infty} C^n v = u$ for some u, depending only

on C and not on v) for any positive matrix C whose column

sums were all unity.

Exercises

10.   If M is any k × k complex matrix, show that

   a. $|M| \leq \max\limits_{j} \sum |m_{ij}|$. (Suggestion: If Mx = λx then

        $\lambda x_i = \sum\limits_{j} m_{ij} x_j$ for all i; in particular, for an index

        i satisfying $|x_i| = \max\limits_{t} |x_t|$.

   b. $|M| \leq \max\limits_{j} \sum\limits_{i} |m_{ij}|$. (Suggestion: Use (a).)

11.   a.   If C > 0, $\sum\limits_{i=1}^{k} c_{ij} = 1$ for all j, and α > 0, show

        that there exists a positive vector p (depending

        only on C) such that for any vector v if $\sum\limits_{i=1}^{k} v_i = \alpha$,

        then $\lim\limits_{n \to \infty} C^n v = \alpha p$.

   b.   If α isn't necessarily positive, but the other

        hypotheses of (a) hold, to what extend does the

        conclusion of (a) hold?

5.   PRIMITIVE MATRICES

A nonnegative matrix A is said to be *primitive* iff $A^m > 0$

for some m. Thus, e.g., A = $\begin{bmatrix} 0 & 1 & 1 \\ 1 & 0 & 0 \\ 1 & 1 & 1 \end{bmatrix}$ is primitive because

$A^4 > 0$. Of course, every positive matrix is primitive.

These matrices are of interest to us because they also

satisfy the conclusions of Perron's theorem and its corollary,
theorem 2, as we shall now show:

Theorem 3. If A is primitive, then $|A|$ is a simple
eigenvalue having a positive eigenvector, and no other
eigenvalue's modulus is $|A|$.

Proof: $A^m > 0$ for some m. Applying Perron's theorem to
A^m, we have $A^m p = |A^m| p$ for some $p > 0$.

Let

$$\hat{p} = \sum_{i=0}^{m-1} |A|^{-i} A^i p$$

then

$$A\hat{p} = \sum_{i=0}^{m-1} |A|^{-i} A^{i+1} p$$

$$= \sum_{i=0}^{m-2} |A|^{-i} A^{i+1} p + |A|^{1-m} A^m p = \sum_{i=1}^{m-1} |A|^{-i+1} A^i p + |A|^{1-m} |A^m| p$$

$$= |A| \left(\sum_{i=1}^{m-1} |A|^{-i} A^i p + p \right) \quad \text{as } |A^n| = |A|^n \text{ for all n (see}$$

$$\text{exercise 13)}$$

$$= |A| \hat{p}$$

As $\hat{p}$ is a sum of positive vectors, it too is positive.
Therefore, $|A|$ is an eigenvalue of A having a positive
eigenvector.

Notice that, in fact, p is also an eigenvector for $|A|$,
because both p and $\hat{p}$ are eigenvectors for the simple eigenvalue
$|A|^m$ of A^m [Exercise 5(a)].

Next we show that $|A|$ is a simple eigenvalue of A: The
multiplicity ℓ of $|A|$ as an eigenvalue of A doesn't exceed

the multiplicity of $|A^m|$ as an eigenvalue of A^m (according
to Exercise 12), but $|A^m|$ is a simple eigenvalue of A^m; so
$\ell = 1$.

Finally, suppose that $|\mu| = |A|$ and μ is an eigenvalue
of A. We have $Ay = \mu y$ for some $y \neq 0$. Therefore, $A^m y = \mu^m y$,
but $|\mu^m| = |\mu|^m = |A|^m = |A^m|$. Since $|A^m| > \nu$ for all
eigenvalues $\nu \neq |A^m|$, we see that $\mu^m = |A^m|$; so $A^m y = |A^m| y$,
but $|A^m|$ is a simple eigenvalue of A^m; so [Exercise 5(a)] its
eigenspace must be spanned by p; consequently, y is a multiple
of p, and hence $\mu = |A|$. []

Theorem 4. If A is primitive, p is any positive
eigenvector for A corresponding to $|A|$, and q is any positive
eigenvector for A^{tr} corresponding to $|A|$, then $\lim_{n\to\infty}(|A|^{-1}A)^n = (q^{tr}p)^{-1}pq^{tr}$.

Proof: Theorem 3 implies that A satisfies the hypotheses
of Exercise 8. []

Exercises

12. If M is any square matrix, λ is any eigenvalue of M of
     multiplicity ℓ, and $n \geq 1$, show that
     a. λ^n is an eigenvalue of M^n
     b. the multiplicity of λ^n as an eigenvalue of M^n is at
         least ℓ.
13. If M is any square matrix and $n \geq 1$, show that $|M^n| = |M|^n$.
14. If A is primitive and λ is any eigenvalue having a
     positive eigenvector, show that $\lambda = |A|$.

6. A TEST FOR PRIMITIVITY

If A is a primitive k × k matrix it turns out that
$A^{(k-1)^2+1} > 0$. (This is not so easy to prove without
developing the theory of nonnegative matrices further. [See
J.C. Holladay and R.S. Varga (1958, pp. 631-634) or D.
Rosenblatt (1958, pp. 631-634) or Seneta (1973, pp. 49-53),
for proofs. The result is due to H. Wielandt (1950).])
This fact enables us to test for primitivity by seeing if
$A^n > 0$ for some convenient $n \geq (k - 1)^2 + 1$, because $A^n > 0$
for all $n \geq (k - 1)^2 + 1$ when A is primitive. (Notice that
when some power of A is positive, then all subsequent powers
have to be positive.)

Exercise

15. a. Is $\begin{bmatrix} 0 & 1 & 0 & 0 & 0 \\ 0 & 0 & 1 & 0 & 0 \\ 0 & 0 & 0 & 1 & 0 \\ 0 & 0 & 0 & 0 & 1 \\ 1 & 1 & 0 & 0 & 0 \end{bmatrix}$ primitive?

     b. Is $\begin{bmatrix} 0 & 0 & 0 & 0 & 1 \\ 1 & 0 & 0 & 0 & 0 \\ 0 & 0 & 1 & 1 & 0 \\ 0 & 0 & 1 & 0 & 0 \\ 1 & 1 & 0 & 0 & 0 \end{bmatrix}$ primitive?

7. PRIMITIVE MATRICES: APPLICATIONS

An association of five teams are to compete in a "round-robin"
tournament. That is, each team plays each of the other teams
exactly once. Their past performance is given by a matrix of
frequencies f_{ij} with which team i has beaten team j in the past:

$$F = \begin{bmatrix} 0 & 0.6 & 0.7 & 0.5 & 0 \\ 0.4 & 0 & 0 & 0.1 & 0.3 \\ 0.3 & 1 & 0 & 0.7 & 0.2 \\ 0.5 & 0.9 & 0.3 & 0 & 0.2 \\ 1 & 0.7 & 0.8 & 0.8 & 0 \end{bmatrix}$$

Thus the third team has won 70% of its games against the fourth team.

We want to arrange a "uniform" system of bets between the teams. That is, team j pays x_j dollars to any team beating it. We also would like the betting scheme to be "fair" - i.e., no team's expected net winnings ought to be negative. Can we do this? Let's analyze the problem. We have $\sum_{i=1}^{k} f_{ij} x_j$ = expected losses of team j, and $\sum_{i=1}^{k} f_{ji} x_i$ = expected winnings of team j. For fairness we require

$$\sum_i f_{ji} x_i = \sum_i f_{ij} x_j \tag{1}$$

for all j for some positive vector x. Let $a_{ji} = f_{ji}/\gamma_j$ where $\gamma_j = \sum_i f_{ij}$. The matrix-vector form of (1) is

$$Ax = x \quad \text{for some } x > 0$$

Exercises

16. Show that A is primitive.
17. Show that $|A| = 1$. (This can be done without examinating the characteristic polynomial of A.)

It follows that our problem has a solution and, apart from scalar multiples, there is only one solution.

Exercise

18. Find the system of bets (to the nearest dollar) for the
    given problem if the largest bet must be $100. You
    might want to do this by computer.

    Notice that all we needed to check to be sure that the
problem had a solution was that A is primitive and $|A| = 1$.

Exercise

19. Suppose the same association of 5 teams wanted to
    arrange prizes for the winners to be paid by the
    association from entry fees paid by the teams. They
    would like to award the same amount of prize money, say
    x_j dollars to each team that beats team j. They would
    also like to arrange the prizes so that the entry fees
    paid by the 5 teams are proportional to the prize value
    of the teams. That is, the ratio of the fee paid by
    team j to the prize x_j awarded each team for defeating
    team j, ought to be the same for all teams so that
    stronger, highly-valued teams pay a higher fee than the
    presumably weaker, lower-valued teams. Again for the
    sake of fairness the association wants no team to
    expect a net loss.
    a.[†]If the total of prize money to be distributed is to
       be $1,000, find a system of prizes x = [x_1, x_2, x_3,
       x_4, x_5] and the corresponding entry fees subject to

[†]A computer would be helpful here.

the restrictions described above. Is the system you

found the only one?

b.   Would there be a solution to this problem for an

arbitrary matrix F of frequencies? If not, what

conditions would have to be placed on F?

8. A GENERALIZATION OF PERRON'S THEOREM

If A isn't primitive we cannot expect all the conclusions of

Perron's theorem to hold. For example, if $A = \begin{bmatrix} 0 & 2 & 0 & 0 \\ 2 & 0 & 0 & 0 \\ 0 & 0 & 2 & 0 \\ 0 & 0 & 0 & 1 \end{bmatrix}$,

then $|A| = 2$, but the only eigenvectors for 2 are of the form
$[\sigma, \sigma, \tau, 0]^{tr}$, none of which is positive. Thus, although
$|A|$ is an eigenvalue, it doesn't have a *positive* eigenvector,
it isn't simple, and, as -2 is also an eigenvalue, we cannot
conclude that $|A| > |\mu|$ for all other eigenvalues μ of A. But
A did have $|A|$ for an eigenvalue, and $|A|$ did have nonnegative
eigenvectors. Will that always happen? The answer is yes.

Theorem 5.   If $A \geq 0$, then $|A|$ is an eigenvalue of A
having a nonnegative eigenvector.

Proof: Let $B_n = A + (1/n)E$, where each $e_{ij} = 1$.

$$0 \leq A < B_{n+1} < B_n \quad \text{for all } n$$

Therefore, $|A| \leq |B_{n+1}| \leq |B_n|$ for all n (see Exercise 20).
Let $\lambda = glb\{|B_n|: n \geq 1\}$, then $|A| \leq \lambda = \lim_{n \to \infty} |B_n|$. Applying
Perron's theorem to each B_n, we obtain positive vectors $y^{(n)}$
such that

$$B_n y^{(n)} = |B_n| y^{(n)} \quad \text{for all } n > 1$$

If we let $z^{(n)} = \left(\sum_{i=1}^{k} y_i^{(n)} \right)^{-1} y^{(n)}$, then for all n

$$B_n z^{(n)} = |B_n| z^{(n)} \qquad z^{(n)} > 0$$

and

$$\sum_{i=1}^{k} z_i^{(n)} = 1$$

As $S = \left\{ z \geq 0: \sum_{i=1}^{k} z_i = 1 \right\}$ is a compact subset of $\underline{R}^k$, there is a subsequence of the $z^{(n)}$ converging to some $z \in S$. [See e.g., Fulks (1961, p. 191)]. Therefore

$$\lim_{m \to \infty} z^{(n_m)} = z$$

$$\lim_{m \to \infty} B_{n_m} z^{(n_m)} = \lim_{m \to \infty} B_{n_m} z = Az$$

$$\lambda z = \lim_{m \to \infty} |B_{n_m}| \lim_{m \to \infty} z^{(n_m)} = \lim_{m \to \infty} |B_{n_m}| z^{(n_m)} = \lim_{m \to \infty} B_{n_m} z^{(n_m)} = Az$$

But $z \neq 0$ since $0 \notin S$, so λ is an eigenvalue of A having a nonnegative eigenvector z. $|\lambda| \leq |A|$ because λ is an eigenvalue of A. On the other hand, $|A| \leq \lambda$, as we saw before. Therefore, $\lambda = |A|$. []

Exercise

20. If $0 \leq A < C$, show that $|A| \leq |C|$.

   (Suggestion: Explain why $|C|^{-1}C$ is power-convergent. Show that $|C|^{-1}A$ is power-bounded, and hence that

$$1 \geq ||C|^{-1}A| = |C|^{-1}|A|.)$$

9. (HOMOGENEOUS) MARKOV CHAINS AND STOCHASTIC MATRICES

Suppose a physical system can be in exactly one of k states
s_1, s_2, ..., s_k at any given time t_n ($t_0 < t_1 < \cdots < t_n < \cdots$)
and that the probability p_{ij} that it is in state s_j at time
t_{n+1}, given it was in state s_i at time t_n, depends (at most)
on i and j and not on n. Suppose that for each i = 1, 2, ..., k
we are also given the probability $p_i^{(0)}$ that the system is in
state s_i initially (at time t_0). If P = $[p_{ij}]$ and $p^{(0)}$ =
$[p_1^{(0)}, p_2^{(0)}, \ldots, p_k^{(0)}]$, then the jth entry in $p^{(0)}P$ is the
probability that the system will be in state s_j at time t_1,
and more generally, the jth entry in the (row) vector $p^{(n)}$ =
$p^{(0)}P^n$ is the probability that the system is in s_j at time
t_n. Thus $p^{(n)}$ describes the condition of the system at time
t_n. Such systems are called (homogeneous) *Markov chains*;
$p^{(0)}$ is called the *initial (distribution) vector* and P is
called the *transition matrix* of the chain. It is customary
to write vectors as row vectors in this context, so just for
this section, that's what we'll do.

The vector $p^{(0)}$ is a typical *stochastic vector*, that is,
$p^{(0)} \geq 0$ and $\sum_{i=1}^{k} p_i^{(0)}$ = 1. The matrix P is a typical
stochastic matrix, i.e., each of its rows is a stochastic
vector.

Example 1: A rat is placed in the chamber marked "3"
of a maze consisting of four chambers connected by runways
between the exits of the chambers as indicated in the diagram:

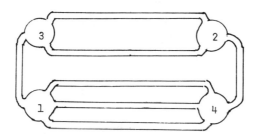

Let us assume that the rat chooses a particular exit or chooses
to stay in the chamber (in a given time interval) with the
same probability. This gives us a transition matrix

$$P = \begin{bmatrix} \frac{1}{5} & 0 & \frac{1}{5} & \frac{3}{5} \\ 0 & \frac{1}{4} & \frac{1}{2} & \frac{1}{4} \\ \frac{1}{4} & \frac{1}{2} & \frac{1}{4} & 0 \\ \frac{3}{5} & \frac{1}{5} & 0 & \frac{1}{5} \end{bmatrix}$$ and initial vector $p^{(0)} = [0,0,1,0]$. For

example: $p_{14} = \frac{3}{5}$ because if the rat is in chember 1 at time
t_n, of five equally likely events (choose exit to 3, choose
first exit to 4, choose second exit to 4, choose third exit
to 4, stay in chamber 1) exactly three events will result in
the rat going to chamber 4; $p_{42} = \frac{1}{5}$ because of the five
equally likely exit choices of a rat in chamber 4 at time t_n
(what are the choices?) only one will result in the rat
being in chamber 2 at time t_{n+1}; but $p_{24} = \frac{1}{4}$ because if the
rat is in chamber 2 at time t_n, then of the four equally
likely exit choices only one will result in him being in
chamber 4 at time t_{n+1}.

   Example 2: This time assume the rat is given a stimulus
at each time t_n which drives it from the chamber it is in.
The transition matrix becomes

$$P = \begin{bmatrix} 0 & 0 & \frac{1}{4} & \frac{3}{4} \\ 0 & 0 & \frac{2}{3} & \frac{1}{3} \\ \frac{1}{3} & \frac{2}{3} & 0 & 0 \\ \frac{3}{4} & \frac{1}{4} & 0 & 0 \end{bmatrix}$$

For example, the probability that a rat in chamber 1 at time t_n chooses an exit leading to chamber 4 is $\frac{3}{4}$, because he has only four equally likely choices (the exit to chamber 3, the first exit to chamber 4, the second exit to chamber 4, and the third exit to chamber 4).

Example 3: *(The game of "craps")*. Let me explain (for the benefit of the one or two of you who have never played it), how this ancient and popular pastime is played. There are at most two stages of the game. In the first stage the player rolls a pair of dice. If he rolls a 7 or an 11 he wins and the game is over. If he rolls 2, 3 or 12 he loses and the game is over. If he rolls any other number (called his "point"), then the game moves into its second stage: The player rolls the dice again and again until he either rolls his point (thus winning the game) or a 7 (thereby losing the game).

We can think of this game as a Markov chain by taking as our states the eight possibilities at the end of stage one:

s_1: player wins

s_2: player loses

s_3: player's point is 4

s_4: player's point is 10

s_5: player's point is 5

s_6: player's point is 9

s_7: player's point is 6

s_8: player's point is 8

The transition matrix is

$$P = \frac{1}{36} \begin{bmatrix} 36 & 0 & 0 & 0 & 0 & 0 & 0 & 0 \\ 0 & 36 & 0 & 0 & 0 & 0 & 0 & 0 \\ 3 & 6 & 27 & 0 & 0 & 0 & 0 & 0 \\ 3 & 6 & 0 & 27 & 0 & 0 & 0 & 0 \\ 4 & 6 & 0 & 0 & 26 & 0 & 0 & 0 \\ 4 & 6 & 0 & 0 & 0 & 26 & 0 & 0 \\ 5 & 6 & 0 & 0 & 0 & 0 & 25 & 0 \\ 5 & 6 & 0 & 0 & 0 & 0 & 0 & 25 \end{bmatrix}$$

The initial vector is

$$p^{(0)} = \frac{1}{36}[8, \ 4, \ 3, \ 3, \ 4, \ 4, \ 5, \ 5]$$

Letting $\underline{1}$ denote $[1, 1, \ldots, 1]^{tr}$, we can say that P is stochastic iff $P \geq 0$ and $P\underline{1} = \underline{1}$ (and p is stochastic iff $p \geq 0$ and $p\underline{1} = 1$). Therefore, 1 is an eigenvalue of P, and hence $|P| \geq 1$, but each row sum of P is 1; so (Exercise 10) $1 \geq |P|$, and consequently *the spectral radius of every stochastic matrix is* 1.

Regular Chains. A chain is *regular* iff no matter what state the system was in initially, there is a time when the system can be in each state with a positive probability. That is, there is a time t_m when $p^{(m)} > 0$, no matter how $p^{(0)}$ was chosen. This is equivalent to requiring that P be primitive. Example 1 above is a regular chain. Examples 2 and 3 are not regular. Applying theorem 4 we see that there is a positive stochastic (row) vector q such that $\lim_{n \to \infty} P^n = \underline{1}q$ and hence, no matter what the initial distribution $p^{(0)}$ was, we must have

$$\lim_{n \to \infty} p^{(n)} = q$$

Therefore when a chain is regular,

    1. The vectors $p^{(n)}$, describing the condition of the system at time t_n, converge as n → ∞.

    2. Although the sequence $p^{(0)}$, $p^{(1)}$, $p^{(2)}$, ..., $p^{(n)}$, ... depends on the choice of $p^{(0)}$ (as $p^{(n)} = p^{(0)}P^n$), the limit of the sequence will be the same vector q regardless of the choice of $p^{(0)}$.

Exercise

21. Find or estimate q for the chain of Example 1. Compare $[0,0,1,0]P^8$ with $[1,0,0,0]P^8$ and interpret your result physically.

Ergodic Chains. A chain is *ergodic* iff no matter what state it was in initially, given any state s_j, there is a time when it is possible for the chain to be in state s_j. In other words, for any $p^{(0)}$ and any j there is some m (m usually depends on $p^{(0)}$ and j) such that $p_j^{(m)} > 0$. This is equivalent to requiring that[†] $\sum_{n=1}^{m} P^n > 0$ for some m. Notice that every regular chain is ergodic. The chain of Example 2 is ergodic. The chain of Example 3 isn't ergodic because,

---

[†] We say P is *ergodic* in this case.

e.g., if the player wins in stage 1 (i.e., the system is in state s_1 at time t_0), then the system will never be in any other state.

Exercise

22. If $\sum_{n=1}^{m} P^n > 0$ for some m, show that $\sum_{n=1}^{k} P^n > 0$.

   (Hint in Exercise 29(b).)

It can be shown (see Sec. 10) that the states of an ergodic chain can be labelled in such a way that the transition matrix P is given by

$$
\begin{bmatrix}
0 & P_1 & 0 & 0 & \cdots & & 0 \\
0 & 0 & P_2 & 0 & \cdots & & 0 \\
\vdots & & & \ddots & & & \vdots \\
& & & & \ddots & & 0 \\
0 & 0 & \cdots & 0 & & & P_{d-1} \\
P_d & 0 & \cdots & 0 & & & 0
\end{bmatrix}
$$

where the blocks of zeros on the diagonal are square (perhaps of different sizes) and P^d is a direct sum of *primitive* stochastic matrices $Q_1 \oplus Q_2 \oplus \cdots \oplus Q_d$. This ensures that

$$\lim_{n \to \infty} P^{nd} = L_1 \oplus L_2 \oplus \cdots \oplus L_d$$

where each L_i is a positive stochastic matrix all of whose rows are the same stochastic vector q_i. Consequently, $P_j L_i = L_i^{(j)}$, a matrix having as many rows as P_j, all of them equal to q_i. Therefore, $P_j L_i^{(\ell)} = L_i^{(j)}$ and we have

$$\lim_{n\to\infty} P^{nd+1} = \begin{bmatrix} 0 & L_2^{(1)} & 0 & \cdots & & 0 \\ 0 & 0 & L_3^{(2)} & 0 & \cdots & 0 \\ \vdots & & \ddots & \ddots & \ddots & \\ \vdots & & & \ddots & \ddots & 0 \\ 0 & 0 & \cdots\cdots & 0 & \ddots & L_d^{(d-1)} \\ L_1^{(d)} & 0 & \cdots\cdots & & 0 & 0 \end{bmatrix}$$

$$\lim_{n\to\infty} P^{nd+2} = \begin{bmatrix} 0 & 0 & L_3^{(1)} & 0 & 0 & \cdots & 0 \\ 0 & 0 & 0 & L_4^{(2)} & 0 & \cdots & 0 \\ \vdots & \vdots & & \ddots & \ddots & \ddots & \\ \vdots & \vdots & & & \ddots & \ddots & 0 \\ 0 & 0 & 0 & \cdots\cdots & 0 & \ddots & L_d^{(d-2)} \\ L_1^{(d-1)} & 0 & 0 & \cdots\cdots & 0 & & 0 \\ 0 & L_2^{(d)} & 0 & \cdots\cdots & 0 & & 0 \end{bmatrix}$$

and

$$\lim_{n\to\infty} P^{nd+d-1} = \begin{bmatrix} 0 & 0 & 0 & \cdots & 0 & L_d^{(1)} \\ L_1^{(2)} & 0 & 0 & \cdots & 0 & 0 \\ 0 & L_2^{(3)} & 0 & \cdots & 0 & 0 \\ \cdot & \cdot & \cdot & \cdot & \cdot & \cdot \\ 0 & 0 & 0 & \cdots & 0 & 0 \\ 0 & 0 & 0 & \cdots & L_{d-1}^{(d)} & 0 \end{bmatrix}$$

Since the L_i are easily found by the methods discussed before for primitive matrices, the vector $p^{(\ell)}$ describing the condition of the system at time t_ℓ is approximated by

$$\begin{bmatrix} 0 & \cdots\cdots\cdots & 0 & L_{r+1}^{(1)} & 0 & 0 & \cdots & 0 \\ 0 & \cdots\cdots\cdots & 0 & 0 & L_{r+2}^{(2)} & 0 & \cdots & 0 \\ & \cdots\cdots\cdots\cdots\cdots\cdots\cdots\cdots & & & & & & 0 \\ L_1^{(d-r+1)} & & & & & & & \vdots \\ 0 & L_2^{(d-r+2)} & \ddots & & & & & \vdots \\ \vdots & & \ddots & & & & & \vdots \\ 0 & 0 & 0\ L_r^{(d)} & 0 & \cdots\cdots\cdots\cdots & & & 0 \end{bmatrix}$$

where r is the remainder when you divide ℓ by d.

 Example 4 (An Ergodic Chain): Suppose a seven state chain (states u, v, w, x, y, z, ℓ) has transition probabilities as given in the diagram:

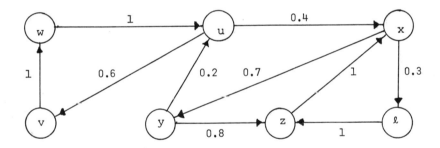

(For example, the probability that the chain is in state v at time t_{n+1} given it was in state u at time t_n is 0.6.) You can verify[†] that no matter what state the chain was in initially, given any state s, there is a path from the initial state to state s, and hence a time when there is a positive probability of being in state s. This makes the chain ergodic.

 If we let s_1 = x, s_2 = v, s_3 = ℓ, s_4 = y, s_5 = w, s_6 = u

---

[†] Consider the circuit going from u to v to w to u to x to y to z to x to ℓ to z to x to y to u.

and $s_7 = \ell$, then we have

$$
P = \left[\begin{array}{cc|ccc|cc}
0 & 0 & 0.3 & 0.7 & 0 & 0 & 0 \\
0 & 0 & 0 & 0 & 1 & 0 & 0 \\
\hline
0 & 0 & 0 & 0 & 0 & 0 & 1 \\
0 & 0 & 0 & 0 & 0 & 0.2 & 0.8 \\
0 & 0 & 0 & 0 & 0 & 1 & 0 \\
\hline
0.4 & 0.6 & 0 & 0 & 0 & 1 & 0 \\
1 & 0 & 0 & 0 & 0 & 0 & 0
\end{array}\right]
$$

$$
P^2 = \left[\begin{array}{cc|ccc|cc}
0 & 0 & 0 & 0 & 0 & 0.14 & 0.86 \\
0 & 0 & 0 & 0 & 0 & 1 & 0 \\
\hline
1 & 0 & 0 & 0 & 0 & 0 & 0 \\
0.88 & 0.12 & 0 & 0 & 0 & 0 & 0 \\
0.4 & 0.6 & 0 & 0 & 0 & 0 & 0 \\
\hline
0 & 0 & 0.12 & 0.28 & 0.6 & 0 & 0 \\
0 & 0 & 0.3 & 0.7 & 0 & 0 & 0
\end{array}\right]
$$

and

$$
P^3 = \left[\begin{array}{cc|ccc|cc}
0.916 & 0.084 & 0 & 0 & 0 & 0 & 0 \\
0.4 & 0.60 & 0 & 0 & 0 & 0 & 0 \\
\hline
0 & 0 & 0.3 & 0.7 & 0 & 0 & 0 \\
0 & 0 & 0.264 & 0.616 & 0.12 & 0 & 0 \\
0 & 0 & 0.12 & 0.28 & 0.6 & 0 & 0 \\
\hline
0 & 0 & 0 & 0 & 0 & 0.656 & 0.344 \\
0 & 0 & 0 & 0 & 0 & 0.14 & 0.86
\end{array}\right]
$$

a direct sum of primitive matrices. Notice that the second summand isn't positive.

Because P^d is a direct sum of primitive (stochastic) matrices, it is easy to calculate $\lim_{n\to\infty} P^{dn}$, and hence to calculate $\lim_{n\to\infty} P^{dn+i}$ for each $0 \le i < d$. This enables us to approximate $p^{(m)}$ for any large m.

In Example 4, we find (using theorem 4, $|A| = 1$ in our case and we can take $p = [1, 1, \ldots, 1]^{tr}$)

$$\lim_{n \to \infty} P^{3n} = \frac{1}{121} \begin{bmatrix} 100 & 21 & 0 & 0 & 0 & 0 & 0 \\ 100 & 21 & 0 & 0 & 0 & 0 & 0 \\ 0 & 0 & 30 & 70 & 21 & 0 & 0 \\ 0 & 0 & 30 & 70 & 21 & 0 & 0 \\ 0 & 0 & 30 & 70 & 21 & 0 & 0 \\ 0 & 0 & 0 & 0 & 0 & 86 & 35 \\ 0 & 0 & 0 & 0 & 0 & 86 & 35 \end{bmatrix}$$

and hence

$$\lim_{n \to \infty} P^{3n+1} = \frac{1}{121} \begin{bmatrix} 0 & 0 & 30 & 70 & 21 & 0 & 0 \\ 0 & 0 & 30 & 70 & 21 & 0 & 0 \\ 0 & 0 & 0 & 0 & 0 & 86 & 35 \\ 0 & 0 & 0 & 0 & 0 & 86 & 35 \\ 0 & 0 & 0 & 0 & 0 & 86 & 35 \\ 100 & 21 & 0 & 0 & 0 & 0 & 0 \\ 100 & 21 & 0 & 0 & 0 & 0 & 0 \end{bmatrix}$$

and

$$\lim_{n \to \infty} P^{3n+2} = \frac{1}{121} \begin{bmatrix} 0 & 0 & 0 & 0 & 0 & 86 & 35 \\ 0 & 0 & 0 & 0 & 0 & 86 & 35 \\ 100 & 21 & 0 & 0 & 0 & 0 & 0 \\ 100 & 21 & 0 & 0 & 0 & 0 & 0 \\ 100 & 21 & 0 & 0 & 0 & 0 & 0 \\ 0 & 0 & 30 & 70 & 21 & 0 & 0 \\ 0 & 0 & 30 & 70 & 21 & 0 & 0 \end{bmatrix}$$

Suppose $p^{(0)} = [0.5, 0.3, 0.2, 0, 0, 0, 0]$ and we want to
estimate $p^{(10,756)}$. For some integer n, $10{,}756 = 3n + 1$ so
$p^{(10,756)} = \lim_{n \to \infty} p^{(0)} P^{3n+1} = \left(\frac{1}{121}\right)[0, 0, 24, 56, 16.8, 17.2, 7]$.

Exercises

23. Approximate $p^{(213)}$ and $p^{(214)}$ for Example 2 if

    a. $p^{(0)} = [0, 0, 1, 0]$

    b. $p^{(0)} = [1, 0, 0, 0]$

    c. $p^{(0)} = [\frac{1}{4}, \frac{1}{4}, \frac{1}{4}, \frac{1}{4}]$

24. Suppose $0 \leq A = \begin{bmatrix} 0_s & B \\ C & 0_t \end{bmatrix}$, where 0_s, 0_t are $s \times s$ and

t × t zero matrices and that BC and CB are primitive

matrices. Show that:

a. $|BC| = |CB| = |A|^2$. (*Hint:* Use theorem 5 on A^2 and

Exercise 14 on BC and CB.

b. $(|A|^{-1}A)^2$ is power convergent and $\lim_{n \to \infty}(|A|^{-1}A)^{2n}$ has

rank 2.

c. $\lim_{n \to \infty}(|A|^{-1}A)^{2n+1} = ?$

Absorbing Chains. A state is said to be *absorbing* iff

once entered it can never be left. A chain is said to be

absorbing iff starting at any state it is possible that

eventually the system will be in an absorbing state. This is

equivalent to requiring that if s_i is arbitrary, then there

is an absorbing s_j possibly depending on i and a time t_m

depending on i and j such that $p_{ij}^{(m)} > 0$. If an absorbing

chain has exactly ℓ absorbing states, one can label the

states so that the transition matrix is $P = \begin{bmatrix} I_\ell & 0 \\ U & T \end{bmatrix}$. (Label

the ℓ absorbing states by s_1, s_2, ..., s_ℓ and the $k - \ell$

nonabsorbing states by $s_{\ell+1}$, ..., s_k.)

$$P^n = \begin{bmatrix} I & 0 \\ \sum_{m=0}^{n-1} T^m U & T^n \end{bmatrix} \quad \text{for all } n \geq 1$$

If $p_{ij}^{(m)} > 0$ and s_j is an absorbing state, then

$$p_{ij}^{(m+r)} \geq p_{ij}^{(m)} p_{jj}^{(r)} = p_{ij}^{(m)} > 0 \quad \text{for all } r$$

Our chain being an absorbing chain, we know that there is an absorbing state s_{j_i} and an integer m_i such that $p_{ij_i}^{(m_i+r)} > 0$

for all $r \geq 0$. Therefore $p_{ij_i}^{(n)} > 0$ for all $i > \ell$ and all

$n \geq \mu = \max_{k>i>\ell} m_i$. Consequently, each row sum of T^μ is less

than 1, and hence (Exercise 10) $|T^\mu| < 1$. This implies that

$|T| < 1$ from which it follows (Chap. 1, Exercise 73) that

$$\lim_{n\to\infty} P^n = \begin{bmatrix} I_\ell & 0 \\ (I - T)^{-1}U & 0 \end{bmatrix}$$

This makes the vectors $p^{(n)}$ converge, but in this case (as opposed to the case of a regular chain) the limit may depend on the choice of $p^{(0)}$. Example 3 illustrates an absorbing chain.

Exercises

25. a. Find $\lim_{n\to\infty} p^{(n)}$ using $p^{(0)}$ and P of Example 3.

     b. What is the probability that the player wins?

     c. If we change the rules a bit so that at stage one
         (only) the player wins, instead of loses, if he
         rolls a 2, find the new initial vector $p^{(0)}$ and the
         corresponding limit vector $\lim_{n\to\infty} p^{(n)}$. What is the
         probability that the player wins now?

26. A and B play the following game: A rolls two (honest)
     dice; if they sum to 3, 5, or 6 then A wins \$1; if they
     sum to 2, A wins \$2 (or whatever B has left). If A
     rolls anything else he loses \$1. A continues to roll

the dice until either A or B have no money left. A

begins the game with D dollars (D = 1, 2, or 3), and B

begins with 4-D dollars. Let s_1, s_2, s_3, s_4, s_5

correspond to the event that A has \$4, \$0, \$1, \$2, \$3,

respectively.

a.   Find $\lim_{n \to \infty} P^n$.

b.   If A starts with \$1, find the probability that

     (eventually) A wins.

There are many chains which are neither absorbing nor

ergodic, but one of the results of the next section

(Exercise 35) implies that (after labelling the states

appropriately) a transition matrix of an arbitrary chain

will either be ergodic, or a direct sum of ergodic matrices,

or of the form

$$P = \begin{bmatrix} E & 0 \\ U & T \end{bmatrix}$$

where E is ergodic or the direct sum of ergodic matrices

and $|T| < 1$. (We saw that transition matrices of absorbing

chains were of this form as I_ℓ is a direct sum of ℓ one by

one ergodic matrices.)

10.   INDECOMPOSABLE MATRICES

In this section we will study a class of nonnegative matrices

(containing all the primitive matrices) which satisfy almost

all of the conclusions of Perron's theorem. These "indecomposable"

matrices also have the useful property that it is easy to

predict how their large powers will behave. In Sec. 9 you
encountered ergodic matrices - these are in fact the
indecomposable stochastic matrices.

Before we define indecomposability, we have to discuss
a special type of matrix: A matrix obtained by permuting
the columns of the identity matrix is called a *permutation*
matrix (e.g., $\begin{bmatrix} 0 & 1 & 0 \\ 0 & 0 & 1 \\ 1 & 0 & 0 \end{bmatrix}$ is one of the six 3 × 3 permutation
matrices). If P is such a matrix, then MP is the matrix
obtained from M by permuting its columns in the same way
that the columns of I were permuted to obtain P. In other
words, the operation of permuting the columns of a matrix M
can be effected by first performing the operation on I (to
obtain a matrix P) and then multiplying to obtain MP. Since
the rows of $P^{tr}M$ are the columns of $M^{tr}P$, it follows that
$P^{tr}M$ is the matrix obtained by permuting the rows of M in
the same way the columns of I were permuted to obtain P,
and hence $P^{tr}MP$ is obtained by permuting the rows and columns
of M the same way. We also know that $P^{tr} = P^{-1}$ because the
dot product of any two distinct columns of P (i.e., columns
of I) is 0 while the dot product of any column of P with
itself is 1, so that $P^{tr}P = I$.

To summarize: if M is any k × k matrix and P is a
permutation matrix, then $P^{-1}MP$ is obtained from M by permuting
the rows and columns of M in the same way.

Example 1.   If A = $\begin{bmatrix} 0 & 2 & 1 & 0 \\ 1 & 0 & 0 & 1 \\ 2 & 0 & 0 & 0 \\ 0 & 1 & 0 & 0 \end{bmatrix}$ and P = $\begin{bmatrix} 1 & 0 & 0 & 0 \\ 0 & 0 & 1 & 0 \\ 0 & 0 & 0 & 1 \\ 0 & 1 & 0 & 0 \end{bmatrix}$, then

then $P^{-1}AP = \begin{bmatrix} 0 & 0 & 2 & 1 \\ 0 & 0 & 1 & 0 \\ 1 & 1 & 0 & 0 \\ 2 & 0 & 0 & 0 \end{bmatrix}$.   We can obtain $P^{-1}AP$ by row and

column permutations:

$$\begin{bmatrix} 0 & 2 & 1 & 0 \\ 1 & 0 & 0 & 1 \\ 2 & 0 & 0 & 0 \\ 0 & 1 & 0 & 0 \end{bmatrix} \xrightarrow[\text{permutation}]{\text{column}} \begin{bmatrix} 0 & 0 & 2 & 1 \\ 1 & 1 & 0 & 0 \\ 2 & 0 & 0 & 0 \\ 0 & 0 & 1 & 0 \end{bmatrix} \xrightarrow[\text{permutation}]{\substack{\text{corresponding} \\ \text{row}}} \begin{bmatrix} 0 & 0 & 2 & 1 \\ 0 & 0 & 1 & 0 \\ 1 & 1 & 0 & 0 \\ 2 & 0 & 0 & 0 \end{bmatrix}$$

$$A \xrightarrow{\hspace{3cm}} AP \xrightarrow{\hspace{3cm}} P^{-1}AP$$

A k × k nonnegative matrix is *decomposable* iff k ≥ 2
and, for some permutation matrix P,

$$P^{-1}AP = \begin{bmatrix} C & D \\ 0 & E \end{bmatrix}$$

where C and E are square matrices. All other nonnegative
matrices are said to be *indecomposable*. In particular every
1 × 1 matrix (even 0) is indecomposable.

For example, $\begin{bmatrix} 1 & 2 & 3 \\ 0 & 0 & 0 \\ 4 & 5 & 6 \end{bmatrix}$ is decomposable because

$$\begin{bmatrix} 1 & 0 & 0 \\ 0 & 0 & 1 \\ 0 & 1 & 0 \end{bmatrix} \begin{bmatrix} 1 & 2 & 3 \\ 0 & 0 & 0 \\ 4 & 5 & 6 \end{bmatrix} \begin{bmatrix} 1 & 0 & 0 \\ 0 & 0 & 1 \\ 0 & 1 & 0 \end{bmatrix} = \begin{bmatrix} 1 & 3 & 2 \\ 4 & 5 & 6 \\ 0 & 0 & 0 \end{bmatrix} \quad C = \begin{bmatrix} 1 & 3 \\ 4 & 6 \end{bmatrix} \quad E = [0]$$

On the other hand, $\begin{bmatrix} 0 & 1 \\ 1 & 0 \end{bmatrix}$ is indecomposable because $P^{-1}\begin{bmatrix} 0 & 1 \\ 1 & 0 \end{bmatrix} P = $
$\begin{bmatrix} 0 & 1 \\ 1 & 0 \end{bmatrix} \neq \begin{bmatrix} c & d \\ 0 & e \end{bmatrix}$ for both 2 × 2 permutation matrices P.

It would be difficult to test for decomposability for
a k × k matrix if we were to rely on the definition alone
because we would have to compute and examine $P^{-1}AP$ for k!
matrices P - which would be quite a task, even if k were as
small as 5.

Fortunately, it turns out that there is an easier way.

Theorem 6.   A is indecomposable iff $\dot{A} + A^2 + \cdots + A^k > 0$ when A is k × k.

Proof: (See Exercise 29 below.)

Example 2:   If $A = \begin{bmatrix} 0 & 0 & 0 & 1 \\ 1 & 0 & 0 & 0 \\ 0 & 0 & 1 & 1 \\ 0 & 1 & 0 & 0 \end{bmatrix}$, then $A + A^2 + A_3 + A^4 \not> 0$

so A is decomposable.

Exercises

27. a.   If A is decomposable and all $\alpha_i \geq 0$, show that

$\sum\limits_{m=1}^{n} \alpha_m A^m$ is decomposable.

  b.   Show that all primitive matrices are indecomposable.

28.   Show that the following two statements are true if, and only if, A is indecomposable:

a. $A + A^2 + \cdots + A^n > 0$ for some $n \geq 1$.

b.   For each i, j, there is some m (depending on i and j) such that $a_{ij}^{(m)} > 0$.

[*Hint*:   First show that (a) holds iff (b) holds. Now suppose (b) is false. It follows that for some i, j, $a_{ij}^{(m)} = 0$ for all $m \geq 1$. Let $S = \bigcup\limits_{m>0} \{\ell : a_{\ell j}^{(m)} > 0\}$ and $T = \bigcup\limits_{m>0} \{\ell : a_{i\ell}^{(m)} > 0\}$. Show that $S \cap T = \emptyset$ (Note: $a_{ij}^{(m+n)} \geq a_{i\ell}^{(m)} a_{\ell j}^{(n)}$).

If $S = \emptyset$, then $a_{\ell j} = 0$ for all ℓ; so column j of A is 0, and hence A is decomposable (Why?). If $T = \emptyset$, show that A is decomposable. Next show that $a_{\ell p} = 0$ if $p \in S$ and $\ell \notin S$ (Note: $a_{\ell j}^{(m+1)} \geq a_{\ell p} a_{pj}^{(m)}$). Finally show that if $S = \{\ell_1, \ell_2, \ldots, \ell_s\}$, and the first s columns $\ell_1, \ell_2, \ldots, \ell_s$ of I and the last k - s columns of P are the remaining columns of I, then

$$P^{-1}AP = \begin{bmatrix} C & D \\ 0 & E \end{bmatrix}$$

where C is s × s and $1 \le s < k$.]

29. a. If $A \ge 0$ and $n \ge 1$, show that $a_{ij}^{(n)} > 0$ iff there exist $j_1, j_2, \ldots, j_{n-1}$ such that $a_{ij_1} a_{j_1 j_2} a_{j_2 j_3} \cdots a_{j_{n-1} j} > 0$.

b. Use 28 and 29(a) to show that the k × k matrix A is indecomposable if, and only if, $A + A^2 + \ldots + A^k > 0$. (Suggestion: If $a_{ij}^{(n)} > 0$ and n > k consider $\{j_1, j_2, \ldots, j_{n-1}, j\}$, and show that $a_{ij}^{(n')} > 0$ for some n' < n.

In 1912 the German mathematician, G. Frobenius, proved a generalization of Perron's theorem for indecomposable matrices. We shall state it in a different but equivalent form. Part (1) and Corollary 1 are the original form of the theorem.

Theorem (Frobenius 1912). If A is an indecomposable k × k matrix and k > 1, then

1. There exists a permutation matrix P and an integer $d \ge 1$ (called the *period* of A) such that:

$$P^{-1}AP = \begin{bmatrix} 0 & 0 & 0 & \cdots & 0 & 0 & A_d \\ A_1 & 0 & 0 & \cdots & 0 & 0 & 0 \\ 0 & A_2 & 0 & \cdots & 0 & 0 & 0 \\ \vdots & \vdots & A_3 & \cdots & 0 & 0 & 0 \\ \vdots & \vdots & \cdot & \cdot & \cdot & \cdot & \cdot \\ 0 & 0 & 0 & \cdots & & 0 & 0 \\ 0 & 0 & 0 & \cdots & 0 & A_{d-1} & 0 \end{bmatrix}$$

where the blocks on the diagonal are square zero
matrices of possibly different sizes, and hence
$P^{-1}A^dP$ is a direct sum of d nonnegative matrices -
call them B_1, B_2, ..., B_d.

2.   The matrices B_i are each primitive, and they all
     have the same spectral radius, namely, $|A|^d$.

Corollary 1.   If A is an indecomposable matrix, then

(i) $|A|$ is a simple eigenvalue having a positive
        eigenvector

(ii)   If d is the period of A, μ is any eigenvalue and
        θ is any d^{th} root of unity, then θμ is also an
        eigenvalue of A.

We'll return to the proofs of Frobenius' theorem and
corollary 1 shortly.

Example 3:   Let A be in Example 1, then

$$P^{-1}AP = \begin{bmatrix} 0 & 0 & 2 & 1 \\ 0 & 0 & 1 & 0 \\ 1 & 1 & 0 & 0 \\ 2 & 0 & 0 & 0 \end{bmatrix} \tag{1}$$

and

$$P^{-1}A^2P = \begin{bmatrix} 4 & 2 & 0 & 0 \\ 1 & 1 & 0 & 0 \\ 0 & 0 & 3 & 1 \\ 0 & 0 & 4 & 2 \end{bmatrix} \tag{2}$$

that is $B_1 = \begin{bmatrix} 4 & 2 \\ 1 & 1 \end{bmatrix}$ and $B_2 = \begin{bmatrix} 3 & 1 \\ 4 & 2 \end{bmatrix}$ (the B_i need not be positive, as they are in this example, but they do have to be primitive in general). We also have $|B_1| = |B_2| = \dfrac{5 + \sqrt{17}}{2}$ $(= |A|^2)$.

At this point you might be curious about how d and P are found. The method of doing so is embedded in a proof of Frobenius' theorem which we'll present shortly; an illustration is given in Example 5 which follows the proof.

The following example illustrates how Frobenius' theorem enables us to estimate $(|A|^{-1}A)^n$ for large n.

Example 4: Suppose A is indecomposable and d = 3. Therefore

$$P^{-1}AP = \begin{bmatrix} 0 & 0 & A_3 \\ A_1 & 0 & 0 \\ 0 & A_2 & 0 \end{bmatrix} \qquad (1)$$

and hence

$$P^{-1}A^3P = \begin{bmatrix} B_1 & 0 & 0 \\ 0 & B_2 & 0 \\ 0 & 0 & B_3 \end{bmatrix} \qquad \begin{matrix} B_1 = A_3A_2A_1 \\ B_2 = A_1A_3A_2 \\ B_3 = A_2A_1A_3 \end{matrix} \qquad (2)$$

and, according to Frobenius' theorem, the B_i are primitive and $|B_i| = |A|^3$ for each i. If we wanted to estimate A^{1048}, we would use (2) to obtain

$$A^{1047} = (A^3)^{349} = P \left[\bigoplus_{i=1}^{3} B_i^{349} \right] P^{-1}$$

Now according to theorem 4, $\lim_{n \to \infty}(|A|^{-3}B_i)^n$ exists, call it L_i. It follows that

$$L_2 = \lim_{n\to\infty}(|A|^{-3}B_2)^n = |A|^{-3}A_1L_1A_3A_2$$

and

$$L_3 = \lim_{n\to\infty}(|A|^{-3}B_3)^n = |A|^{-3}A_2A_1L_1A_3$$

We've seen how to calculate (or estimate) L_1, this one calculation (or estimation) provides us with L_2 and L_3, and A^{1047} is approximately $|A|^{1047} P\left(\bigoplus_{i=1}^{3} L_i\right)P^{-1}$, and hence $A^{1048} = A|A|^{1047} P\left(\bigoplus_{i=1}^{3} L_i\right)P^{-1}$, approximately.

We can formalize the general statement embodied in Example 4 by the following.

Corollary 2. Suppose A is indecomposable, d, P, and B_i are as in Frobenius' Theorem and $L_i = \lim_{n\to\infty} B_i^n$. We then have (1) $\lim_{m\to\infty}|\Lambda|^{-1}A^{dm} = P\left(\bigoplus_{i=1}^{d} L_i\right)P^{-1}$ and, consequently, (2) if n is large and r is the remainder when n is divided by d, then $(|A|^{-1}A)^n$ is approximately $|A|^{-r}A^r P\left(\bigoplus_{i=1}^{d} L_i\right)P^{-1}$.

We have seen that once one of the L_i (say L_1) is evaluated, then the others can be easily computed from it. In this way a large power of A can be calculated by means of a few products of the A_i and L_1.

Exercise

30. Suppose $A = \begin{bmatrix} 0 & 0 & A_3 \\ A_1 & 0 & 0 \\ 0 & A_2 & 0 \end{bmatrix}$, $A_1 = \begin{bmatrix} 1 & 1 \\ 1 & 0 \end{bmatrix}$, $A_2 = \begin{bmatrix} 0 & 1 \\ 1 & 1 \\ 1 & 0 \end{bmatrix}$, and $A_3 = \begin{bmatrix} 1 & 1 & 0 \\ 0 & 0 & 1 \end{bmatrix}$.

a. Find B_1.

b. Show that $|B_1| = 2 + \sqrt{2}$.

c. Show that $L_1 = \lim_{n\to\infty}(|B_1|B_1)^n = (\tfrac{1}{4})\begin{bmatrix} 2 + \sqrt{2} & \sqrt{2} \\ \sqrt{2} & 2 - \sqrt{2} \end{bmatrix}$.

   (Suggestion: Use theorem 2.)

d. Without calculating any further limits, show that

$$L_2 = (\tfrac{1}{4})\begin{bmatrix} 2 & 2\sqrt{2} \\ \sqrt{2} & 2 \end{bmatrix} \quad L_3 = (\tfrac{1}{4})\begin{bmatrix} 1 & 1 & \sqrt{2} - 1 \\ 1 + \sqrt{2} & 1 + \sqrt{2} & 1 \\ \sqrt{2} & \sqrt{2} & 2 - \sqrt{2} \end{bmatrix}$$

e. Estimate A^{2037}.

The proof of Frobenius' theorem which we will give is essentially due to V. Romanovsky (1938). It depends on Perron's theorem and a little theorem of number theory[†]. You may recall that an integer n (positive or otherwise) is said to be a *divisor* of an integer m, and m is said to be a *multiple* of n, iff nq = m for some integer q. If S is any set of positive integers (even an infinite set), an integer c is said to be a *common divisor* of S iff c is a divisor of each element of S. There are only finitely many of these - the largest is called the *greatest common divisor of* S and is denoted by gcd(S).

   Lemma 10.1: If S is a set of positive integers which is closed under addition (that is, x + y is in S whenever x and y are in S), then every sufficiently large multiple of gcd(S) is in S.

   Proof: See Exercise 34.

[†] The number-theoretic part of the proof is partially based on Chung (1960), pp. 11-13.

A Proof of Frobenius' theorem: Let $S = \{n > 0 : a_{11}^{(n)} > 0\}$ and $d = \gcd S$. S is closed under addition (because $a_{11}^{(n+m)} \geq a_{11}^{(n)} a_{11}^{(m)}$). Therefore, by Lemma 10.1, every sufficiently large multiple of d lies in S, i.e.,

1. $a_{11}^{(dn)} > 0$ for all sufficiently large n.

Next we prove an extension of (1):

2. For each $j \geq 1$ there is an integer r_j $(0 \leq r_j < d)$ such that if $a_{j1}^{(m)} > 0$, then $m = dn_0 + r_j$ for some $n_0 \geq 0$, and $a_{j1}^{(dn+r_j)} > 0$ for all sufficiently large n.

Fix j. Exercise 28 implies that $a_{j1}^{(t)} > 0$ and $a_{1j}^{(\ell)} > 0$ for some t and ℓ. Let r_j be the remainder when t is divided by d. Now for any n: $a_{11}^{(n+\ell)} \geq a_{1j}^{(\ell)} a_{j1}^{(n)}$. Therefore if $a_{j1}^{(n)} > 0$, then $n + \ell$ is a multiple of d. So if $a_{j1}^{(m)} > 0$, then $m + \ell$ and $t + \ell$ are both multiples of d, and hence $m = t + dq$ for some $q \geq 0$. But $t = dp + r_j$ for some $p \geq 0$, therefore $m = d(p+q) + r_j$. Conversely, if $m = dn + r_j$ and $n \geq p$ then $m = d(n - p) + t$ but $a_{j1}^{(m)} \geq a_{j1}^{(t)} a_{11}^{(d(n - p))}$ and hence $a_{j1}^{(m)} > 0$ for all sufficiently large m.

Let $C_r = \bigcup_{n \geq 0} \{j : a_{j1}^{(dn+r)} > 0\}$ for $r = 0, 1, \ldots, d-1$. No $C_r = \emptyset$, otherwise the first column of A^r, and hence of A, would be 0 and A would be decomposable. According to (2) we have $\{1, 2, \ldots, k\} = \bigcup_{r=0}^{d-1} C_r$.

The C_r are pairwise disjoint because if $j \in C_r \cap C_s$, then $a_{j1}^{(dq+r)} > 0$ and $a_{j1}^{(dq'+s)} > 0$; so $dq + r = dn + r_j$ and

$dq' + s = dn' + r_j$ by (2), and (since r, s and r_j are between
0 and d - 1) we have $r = r_j = s$.

    3. If $a_{ij} > 0$ and $j \in C_r$, then $i \in C_{r+1}$ for each[†]
        $0 \leq r < d$ (this is because $a_{i1}^{(n+1)} \geq a_{ij} a_{j1}^{(n)}$).

Now we can construct[††]the matrix P which we need to
prove Frobenius' theorem. Suppose we list the elements in
C_r by $i_{r_1}, i_{r_2}, \ldots, i_{r_{c_r}}$ (c_r is the number of members of C_r
and $C_r = \{i_{r_1}, \ldots, i_{r_{c_r}}\}$). Let P be the permutation matrix
whose first c_0 columns are columns $i_{0_1}, i_{0_2}, \ldots, i_{0_{c_0}}$ of I;
whose next c_1 columns are columns $i_{1_1}, i_{1_2}, \ldots, i_{1_{c_1}}$ of I;
etc. Applying (3) we obtain:

$$P^{-1}AP = \begin{bmatrix} 0 & 0 & \cdots & 0 & A_d \\ A_1 & 0 & \cdots & 0 & 0 \\ 0 & A_2 & & & 0 \\ \vdots & \vdots & \ddots & & \\ 0 & 0 & & \ddots\, 0 & 0 \\ 0 & 0 & & A_{d-1} & 0 \end{bmatrix} \qquad (1)$$

where the blocks on the diagonal are $c_r \times c_r$ zero matrices,
the A_r are $c_{r+1} \times c_r$ and the only positive entries in $P^{-1}AP$
are in the A_r (but there may be some zeros in A_r). Consequently

    4. $P^{-1}A^d P = \overset{d}{\underset{i=1}{\oplus}} B_i$
        where $B_1 = A_d A_{d-1} \cdots A_2 A_1$, $B_2 = A_1 A_d \cdots A_3 A_2$, $\ldots$,
        $B_d = A_{d-1} A_{d-2} \cdots A_1 A_d$.

[†] If $j \in C_{d-1}$ then $i \in C_0$.

[††]Example 5 below illustrates this procedure.

Now we'll prove part (2) of the theorem. Suppose $j \in C_0$, then $a_{1j}^{(dn_j)} > 0$ for some n_j because of (4) and the fact that A is indecomposable. But $a_{11}^{(dn)} > 0$ for all sufficiently large n [by (1)]; so $a_{1j}^{(d(n_j+n))} > 0$ for these n. Consequently, there is some n_0 (independent of j) such that $a_{1j}^{(dn_0)} > 0$ for all $j \in C_0$. If $i \in C_0$, then $a_{i1}^{(dn_i)} > 0$ for some n_i, but $a_{11}^{(dn)} > 0$ for all sufficiently large n; so $a_{i1}^{[d(n_i+n)]} > 0$ for these n, and hence there is some m_0 (independent of i) such that $a_{i1}^{(dm_0)} > 0$ for all $i \in C_0$. Therefore $a_{ij}^{d(m_0+n_0)} > 0$ for every i, j $\in C_0$, and hence B_1 is primitive. $B_2^{n+1} = A_1 B_1^n A_d A_{d-1} \cdots A_2$ for all n, no A_m has a zero row or column so $B_2^{n+1} > 0$ if $B_1^n > 0$, and hence B_2 is primitive. Similarly, we can show that $B_3, \ldots, B_d$ are primitive.

$B_1 p = |B_1| p$ for some $p > 0$ and $B_2 q = |B_2| q$ for some $q > 0$ because B_1, B_2 are primitive. Now $A_1 B_1 = B_2 A_1$ and $A_1 B_1 p = |B_1| A_1 p$ so $B_2(A_1 p) = |B_1|(A_1 p)$ and hence $|B_1|$ is a positive eigenvalue for B_2 belonging to the positive eigenvector $A_1 p$. Exercise 14 implies that $|B_2| = |B_1|$. Similarly, it can be shown that $|B_i| = |B_1|$ for all $i \leq d - 1$. Exercise 13 implies that $|B_1| = |A|^d$. So we have shown that the B_i are primitive matrices with a common spectral radius, $|A|^d$, completing the proof of Frobenius' theorem. []

Example 5. (Using the proof of Frobenius' theorem to find P.) Suppose

$$A = \begin{bmatrix} 0 & 0 & 0 & 1 & 0 & 1 & 0 \\ 1 & 0 & 0 & 0 & 1 & 0 & 0 \\ 1 & 0 & 0 & 0 & 0 & 0 & 0 \\ 0 & 0 & 1 & 0 & 0 & 0 & 0 \\ 0 & 0 & 0 & 0 & 0 & 0 & 1 \\ 0 & 1 & 1 & 0 & 0 & 0 & 0 \\ 0 & 1 & 0 & 0 & 0 & 0 & 0 \end{bmatrix}$$

Let u_n denote the first column of A^n -- to compute u_n just
use $u_2 = Au_1$, $u_3 = Au_2$, ..., etc. We'll calculate them one
by one and keep track of which entries are positive in each
u_n be writing "+" in the jth row of u_n to indicate $a_{j1}^{(n)} > 0$.
We stop when the pattern of 0's and +'s repeats. We obtain
in our example

$$
\begin{array}{cccc}
u_1 & u_2 & u_3 & u_4 \\
\begin{bmatrix} 0 \\ + \\ + \\ 0 \\ 0 \\ 0 \\ 0 \end{bmatrix} &
\begin{bmatrix} 0 \\ 0 \\ 0 \\ + \\ 0 \\ + \\ + \end{bmatrix} &
\begin{bmatrix} + \\ 0 \\ 0 \\ 0 \\ + \\ 0 \\ 0 \end{bmatrix} &
\begin{bmatrix} 0 \\ + \\ + \\ 0 \\ 0 \\ 0 \\ 0 \end{bmatrix}
\end{array}
$$

which means that $a_{j1}^{(4)} > 0$ iff $a_{j1}^{(1)} > 0$. Therefore $a_{j1}^{(3n+1)} > 0$
iff $a_{j1}^{(1)} > 0$ for all n, j and hence $a_{j1}^{(3n+2)} > 0$ iff $a_{j1}^{(2)} > 0$
for all n, j, and hence $a_{j1}^{(3n)} > 0$ iff $a_{j1}^{(3)} > 0$ for all n, j.
This tells us that $d = 3$ and $C_1 = \{2, 3\}$, $C_2 = \{4, 6, 7\}$, and
$C_0 = \{1, 5\}$, because $a_{j1}^{(1)} > 0$ iff $j = 2$ or 3, $a_{j1}^{(2)} > 0$ iff
$j = 4$, 6, or 7, and $a_{j1}^{(3)} > 0$ iff $j = 1$ or 5.

Having determined the C_r we apply the method given in
the proof to find P:

$$P = \begin{bmatrix} 1 & 0 & 0 & 0 & 0 & 0 & 0 \\ 0 & 0 & 1 & 0 & 0 & 0 & 0 \\ 0 & 0 & 0 & 1 & 0 & 0 & 0 \\ 0 & 0 & 0 & 0 & 1 & 0 & 0 \\ 0 & 1 & 0 & 0 & 0 & 0 & 0 \\ 0 & 0 & 0 & 0 & 0 & 1 & 0 \\ 0 & 0 & 0 & 0 & 0 & 0 & 1 \end{bmatrix}$$

and hence

$$P^{-1}AP \; (= P^{tr}AP) = \begin{bmatrix} 0 & 0 & 0 & 0 & 1 & 1 & 0 \\ 0 & 0 & 0 & 0 & 0 & 0 & 1 \\ 1 & 1 & 0 & 0 & 0 & 0 & 0 \\ 1 & 0 & 0 & 0 & 0 & 0 & 0 \\ 0 & 0 & 0 & 1 & 0 & 0 & 0 \\ 0 & 0 & 1 & 1 & 0 & 0 & 0 \\ 0 & 0 & 1 & 0 & 0 & 0 & 0 \end{bmatrix}$$

$$= \begin{bmatrix} 0 & 0 & A_3 \\ A_1 & 0 & 0 \\ 0 & A_2 & 0 \end{bmatrix}$$

where $A_1 = \begin{bmatrix} 1 & 1 \\ 1 & 0 \end{bmatrix}$, $A_2 = \begin{bmatrix} 0 & 1 \\ 1 & 1 \\ 1 & 0 \end{bmatrix}$, and $A_3 = \begin{bmatrix} 1 & 1 & 0 \\ 0 & 0 & 1 \end{bmatrix}$. Therefore

$$P^{-1}A^3P = \begin{bmatrix} 3 & 1 \\ 1 & 1 \end{bmatrix} \oplus \begin{bmatrix} 2 & 2 \\ 1 & 2 \end{bmatrix} \oplus \begin{bmatrix} 1 & 1 & 0 \\ 2 & 2 & 1 \\ 1 & 1 & 1 \end{bmatrix}$$

The direct summands are all primitive and have $\mu = 2 + \sqrt{2}$ as their common spectral radius.

Exercises

31. Complete the proof of Frobenius' theorem by showing that:

    a. All the B_i are primitive (we did B_1, B_2 in the proof), and

    b. $|B_i| = |B_1|$ for all i (we did it for $i = 2$ in the proof).

32.   Let A =
$$A = \begin{bmatrix} 0 & 0 & 0 & 0 & 0 & 1 \\ 1 & 0 & 0 & 0 & 0 & 0 \\ 0 & 0 & 0 & 1 & 0 & 1 \\ 0 & 1 & 0 & 0 & 0 & 0 \\ 2 & 0 & 1 & 0 & 0 & 0 \\ 0 & 0 & 0 & 0 & 1 & 0 \end{bmatrix}$$

a.   Find $P^{-1}AP$ as in Frobenius' theorem.

b.   Estimate $(|A|^{-1}A)^{1048}$.

33.   Suppose $\underline{Z}$ denotes the integers (negative and nonnegative), $V = \{s_1, s_2, \ldots, s_p\} \subset \underline{Z}$, d = gcd(V).   If T =

$\{ \sum_{i=1}^{p} z_i s_i : z_i \in \underline{Z} \}$ show that T = $\{zd : z \in \underline{Z}\}$.

(Suggestions: Let m = min$\{t \in T : t > 0\}$ [we can assume T $\neq \{0\}$].

a.   Show that m is a divisor of d (divide $t \in T$ by m; obtain t = mq + r with $0 \leq r < m$; explain why $r \in T$, r = 0, and hence why m is a common divisor of V.

b.   Show that d is a common divisor of T, explain why d is a divisor of m.)

34.   Suppose S is a nonempty subset of the positive integers which is closed under addition and d = gcd(S).

a.   Show that d = gcd(V) for some finite subset V of S.

b.   Suppose V = $\{s_1, s_2, \ldots, s_p\}$, z = $\sum_{i=1}^{p} x_i s_i$ $(x_i \in \underline{Z})$ and z $\geq \left(\sum_{i=1}^{p} s_i \right)^2$.   Show that there exist $y_i \in \underline{Z}$

such that

$$\begin{cases} y_i > x_i & \text{if } x_i < 0 \\ y_i \geq 0 & \text{if } x_i \geq 0 \end{cases}$$

and z = $\sum_{i=1}^{p} y_i s_i$.   (Suggestion: relabel the x_i if necessary so that $x_i < 0$ if $1 \leq i \leq t$ and $x_p \geq x_i \geq 0$

for all $t < i \leq p$. Now $x_p \left(\sum\limits_{i=1}^{p} s_i \right) \geq z \geq \left(\sum\limits_{i=1}^{p} s_i \right)^2$

and hence $x_p \geq \sum\limits_{i=1}^{p} s_i$. On the other hand,

$$z = \sum\limits_{i=1}^{t} s_i (x_i + s_p) + \sum\limits_{i=t+1}^{p-1} x_i s_i + s_p (x_p - \sum\limits_{i=1}^{t} s_i)$$

now choose y_i appropriately.)

c. Prove lemma 10.1 by first using Exercises 34(a) and

   33 and then applying 34(b) several times.

Example 6 (An Application of Corollary 1). Suppose
A is as in Example 5 and we'd like to find all the eigenvalues
of A. The eigenvalues of B_3 are $\mu_1 = 2 + \sqrt{2}$, $\mu_2 = 2 - \sqrt{2}$,
and 0. Let λ_j be the real cube root of μ_j ($j = 1$, 2) and ω
be a complex cube root of 1 [say $\omega = \frac{1}{2}(-1 + 2i\sqrt{3})$], then
according to Exercise 12 and corollary 1(ii): λ_1, $\lambda_1 \omega$, $\lambda_1 \omega^2$,
λ_2, $\lambda_2 \omega$, $\lambda_2 \omega^2$, and 0 are the eigenvalues of A.

   Since there are seven distinct eigenvalues and A is
7 × 7, it follows that A is diagonable. We can also see
immediately that the characteristic polynomial of A must be
$(\tau^3 - \mu_1)(\tau^3 - \mu_2)\tau = [\tau^6 - (\mu_1 + \mu_2)\tau^3 + \mu_1\mu_2]\tau = (\tau^6 - 4\tau^3 + 2)\tau$.
Notice that corollary 1(ii) gives us the following geometric
picture when we plot the eigenvalues of A in the complex plane:

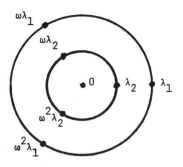

A proof of Corollary 1: (B_1, P, c_i, etc., are as in the proof of Frobenius' theorem.) Let $F = P^{-1}AP$. Since B_1 is primitive, we have $B_1 p = |A|^d p$ for some $p > 0$ in $\underline{R}^{c_0}$. Let

$$u = [p_1, p_2, \ldots, p_{c_0} ; \underbrace{0, 0, \ldots, 0]}_{k-c_0}{}^{tr}$$

and $\lambda = |A|$ ($= |F|$).

If we put $v = \sum_{r=0}^{d-1} \lambda^{-r} F^r u$, then we obtain $Fv = \sum_{r=0}^{d-1} \lambda^{-r} F^{r+1} u = \left(\sum_{r=1}^{d-1} \lambda^{1-r} F^r u\right) + \lambda^{-d}\lambda^d u = \lambda v$, and hence $A(Pv) = |A|(Pv)$; so

Pv is a positive eigenvector for the eigenvalue $|A|$ of A. Part (2) of Frobenius' theorem implies that A is powerbounded so the Jordan blocks corresponding to λ are all 1×1. Suppose $Fx = \lambda x$. For each $z \in \underline{C}^k$, let

$$z^{(0)} = [z_1, z_2, \ldots, z_{c_0} ; \underbrace{0, 0, \ldots, 0]}_{k-c_0}$$

$$z^{(1)} = [\underbrace{0, 0, \ldots 0}_{c_0} ; z_{c_0+1}, \ldots, z_{c_0+c_1}; \underbrace{0, 0, \ldots, 0]}_{k-c_0-c_1}$$

$$\vdots$$

$$z^{(d-1)} = [\underbrace{0, 0, \ldots, 0}_{c_0 + \cdots + c_{d-2}} ; z_{k-c_{d-1}+1}, \ldots, z_k]$$

then $Fx^{(0)} = \lambda x^{(1)}$, $Fx^{(1)} = \lambda x^{(2)}$, $\ldots$, $Fx^{(d-1)} = \lambda x^{(0)}$, thus $x^{(r)} = \lambda^{-r} F^{(r)} x^{(0)}$ for $r = 0, 1, 2, \ldots, d-1$. Therefore

$x = \sum_{r=0}^{d-1} \lambda^{-r} F^{(r)} x^{(0)}$, but $x^{(0)} = \tau u$ for some τ (apply Exercise

5(a) to B_1), and hence $x = \tau v$. Consequently there is only one Jordan block corresponding to λ, since that block is

1×1, it follows that A is a simple eigenvalue of A. This completes (i) of the corollary.

To prove (ii); If $Fw = \mu w$, then $\mu w^{(r+1)} = F(w^{(r)})$ for each $r = 0, 1, \ldots, d-2$, and $\mu w^{(0)} = F(w^{(d-1)})$. Suppose $\theta^d = 1$. Let $h = \sum_{r=0}^{d-1} \theta^{-r} w^{(r)}$, then $Fh = (\theta \mu)h$ by the same sort of calculation we did for (i). Therefore, $\theta \mu$ is an eigenvalue of A if μ is an eigenvalue and θ is a dth root of unity. []

Exercises

(A Normal Form for Nonnegative Matrices)

35. If A is any $k \times k$ matrix, show that for some permutation matrix P:

$$P^{-1}AP = \begin{bmatrix} F_1 & & & & & \\ & F_2 & & & \text{\Large$*$} & \\ & & \ddots & & & \\ & & & F_i & & \\ & \text{\Large$\bigcirc$} & & & \ddots & \\ & & & & & F_s \end{bmatrix}$$

where for each $1 \leq i \leq s$, F_i is an indecomposable matrix. (*Note:* $P^{-1}AP$ is not necessarily a direct sum; there may be positive entries above the F_i.) *Hint:* Proceed by induction on k. If A is decomposable, then $P^{-1}AP = \begin{bmatrix} C & D \\ 0 & E \end{bmatrix}$ for some permutation matrix P, where C is an $s \times s$ matrix with $1 \leq s < k$ and E is $(k - s) \times (k - s)$. Now use the fact that

$$\begin{bmatrix} QGQ^{-1} & D \\ 0 & RHR^{-1} \end{bmatrix} = \begin{bmatrix} Q & 0 \\ 0 & R \end{bmatrix} \begin{bmatrix} G & D \\ 0 & H \end{bmatrix} \begin{bmatrix} Q^{-1} & 0 \\ 0 & R^{-1} \end{bmatrix}$$

36. Prove that any *doubly stochastic* matrix D (i.e., both
    D and D^{tr} are stochastic) is a direct sum of indecomposable
    doubly stochastic matrices. (*Hint:* Suppose $P^{-1}AP =$
    $\begin{bmatrix} F & X \\ 0 & Y \end{bmatrix}$, where F is s × s, then F^{tr} and Y are stochastic.
    (Why?) If P is m × m and stochastic, then $\sum_{i=1}^{m} \sum_{j=1}^{m} P_{ij} = m$.
    Use this fact to evaluate the sum of the entries in $P^{-1}AP$
    in two different ways which will show that the sum of
    the entries in X is zero, and hence X = 0. Then apply
    Exercise 35.

37. If D is doubly stochastic, what can you say about
    $\lim_{n \to \infty} D^n$?

38. Suppose, as in Sec. 7, that M is a k × k nonnegative
    matrix such that $m_{ii} = 0$ for all i and $m_{ij} + m_{ji} = 1$
    for all i ≠ j.

    a. If M is indecomposable and k > 3, show that M is
       primitive.

    b. Show that the teams (as in Sec. 7) can be labeled
       so

    where the M_i are square ($m_i \times m_i$) possibly of
    different sizes, there are only zeros below the
    M_i, only 1's above the M_i, and M_i is primitive if

$m_i > 3$; M_i is primitive or $= \begin{bmatrix} 0 & 1 & 0 \\ 0 & 0 & 1 \\ 1 & 0 & 0 \end{bmatrix}$ if $m_i = 3$;

$M_i = \begin{bmatrix} 0 & 1 - a_i \\ a_i & 0 \end{bmatrix}$ for some $0 < a_i < 1$ if $m_i = 2$; and

$M_i = 0$ if $m_i = 1$. Notice that this labeling

provides a natural way of putting the teams into

leagues (the first m_1 teams in league 1, the next

m_2 teams in league 2, etc.). Since the outcome

between teams in different leagues is certain,

competition can be restricted to take place only

among teams in the same league. [*Hint:* (a) if M

were indecomposable but not primitive, then, by

Frobenius' theorem, $P^{-1}MP = \begin{bmatrix} 0 & U \\ V & W \end{bmatrix}$, where the zero

block is s × s with s > 1. Now use the fact that

every off-diagonal entry in $M + M^{tr}$ is 1. For (b)

use Exercise 35.]

39. a. If A is indecomposable and nonsingular, show that

all the A_i have the same size.

b. If A is a k × k, indecomposable, nonsingular matrix

and k is a prime, show that A is primitive, or

A = PDQ, where D is a diagonal matrix and P, Q are

permutation matrices. (*Hint:* (a) If $B = \begin{bmatrix} 0_s & U \\ V & W \end{bmatrix}$,

where W is t × t, s + t = k, and s > t $\geq$ 1, then

rank(B) = rank(U) + rank([V,W]), rank(U) $\leq$ t, and

rank([V,W]) $\leq$ t (Why?); therefore, rank(B) $\leq$ 2t < k.

CHAPTER 3

DIFFERENTIAL EQUATIONS

1. MATRIX-VALUED FUNCTIONS OF ONE SCALAR VARIABLE

Suppose $m_{11}(t)$, $m_{12}(t)$, ..., $m_{ij}(t)$, ..., $m_{k\ell}(t)$ are scalar valued functions of one scalar variable t. The k × ℓ matrix whose ij^{th} entry is $m_{ij}(t)$ is a matrix-valued function of one scalar variable -- we denote it by M(t).

$$Example \ 1: \quad A(t) = \begin{bmatrix} \sin t & \cos t & t \\ \dfrac{\sin t}{t} & e^t & t^2 \\ 1 & 0 & t^3 \end{bmatrix} \quad (t \neq 0)$$

$$Example \ 2: \quad x(t) = \begin{bmatrix} e^t \\ \dfrac{1}{t^2} \\ 1 \end{bmatrix} \quad (t \neq 0, \ t \ \text{real})$$

$$Example \ 3: \quad y(t) = \begin{bmatrix} \sin \dfrac{\pi}{t} \\ \sin t\pi \end{bmatrix} \quad (t \neq 0)$$

We say that $\lim_{t \to \alpha} M(t) = M$ iff $\lim_{t \to \alpha} m_{ij}(t) = m_{ij}$ for all i, j. We allow $\alpha = \infty$ (or $-\infty$ when t is a real variable) but we require all m_{ij} to be finite. Thus

$$\lim_{t \to 0} A(t) = \begin{bmatrix} 0 & 1 & 0 \\ 1 & 1 & 0 \\ 1 & 0 & 0 \end{bmatrix} \quad \text{for A \ as in Example 1}$$

$$\lim_{t \to -\infty} x(t) = \begin{bmatrix} 0 \\ 0 \\ 1 \end{bmatrix} \quad \text{for x \ as in Example 2}$$

118

and the function y of Example 3 has no limit as t → ∞, and it has no limit as t → 0. We say that the matrix-valued function M(t) is *continuous* at t = α (α finite) iff M has a limit at α and $\lim_{t \to \alpha} M(t) = M(\alpha)$. We say that the matrix-valued function M(t) is *differentiable* at t (t finite) iff $\lim_{s \to t}[1/(s-t)][M(s) - M(t)]$ exists, in which case we call the limit the *derivative of M at t* and denote it by M'(t), Ṁ(t), or dM/dt.

Thus

$$\frac{dA}{dt} = \begin{bmatrix} \cos t & -\sin t & 1 \\ \dfrac{t \cos t - \sin t}{t^2} & e^t & 2t \\ 0 & 0 & 3t^2 \end{bmatrix} \quad \text{for all } t \neq 0$$

when A is as in Example 1 because the ij^{th} entry in dM/dt turns out to be $d(m_{ij})/dt$ (see Exercise 1).

If each of the scalar functions $m_{ij}(t)$ are integrable on the real interval [α, β], we say that M(t) is *integrable*, and we define

$$\int_\alpha^\beta M(t) \, dt = \left[\int_\alpha^\beta m_{ij}(t) \, dt \right]$$

(= the matrix whose ij^{th} entry is $\int_\alpha^\beta m_{ij}(t) \, dt$). Thus

$$\int_0^1 \begin{bmatrix} t \sin \pi t \\ 1 & -t \end{bmatrix} dt = \begin{bmatrix} \dfrac{1}{2} & \dfrac{2}{\pi} \\ 1 & -\dfrac{1}{2} \end{bmatrix}$$

Exercises

1. If M is a k × ℓ matrix-valued function of one scalar
    variable, show that:

    a. M is continuous at t = α iff each $m_{ij}(t)$ is continuous
        at t = α.

    b. M is differentiable at t = α iff each $m_{ij}(t)$ is
        differentiable at t = α.

    c. If M is differentiable at t, then the ij^{th} entry of
        dM/dt is dm_{ij}/dt.

2. Suppose M and N are differentiable for all t ε S and
    α, β are scalars.

    a. Show that αM + βN is differentiable and $\frac{d}{dt}(\alpha M + \beta N) =$
        $\alpha\frac{dM}{dt} + \beta\frac{dN}{dt}$ (for all t ε S).

    b. If M and N are k × ℓ and ℓ × r respectively, then
        MN is also differentiable and $\frac{d}{dt}(MN) = \frac{dM}{dt}N + M\frac{dN}{dt}$
        (all t ε S).

    c. If M is differentiable and invertible at each t ε S,
        then M^{-1} is differentiable and $dM^{-1}/dt = -M^{-1}\frac{dM}{dt}M^{-1}$
        for all t ε S.

3. Suppose M(t) is a matrix-valued function which is continuous
    on an interval [α, β]. Show that $\int_{\alpha}^{t}M(s)\,ds$ is differentiable
    at each t in the interval and that

    $$\frac{d}{dt}\int_{\alpha}^{t} M(s)\,ds = M(t) \quad \text{for all } \alpha < t < \beta$$

4. Suppose M(t) is a matrix-valued function which is
    continuously differentiable on an interval [α, β]. Show

that for all $\alpha \leq t \leq \beta$:

$$\int_{\alpha}^{t} \frac{dM}{ds} \, ds = M(t) - M(\alpha)$$

(Suggestion: Use the corresponding result for m_{ij}.)

2. A MATRIX NORM

There is a convenient matrix analogue of the magnitude (absolute value) of a scalar which we shall make extensive use of in the work ahead. If M is a $k \times \ell$ matrix (this includes the possibility of M being a vector), we define

$$||M|| = \sum_{i=1}^{k} \sum_{j=1}^{\ell} |m_{ij}|$$

and we refer to $||M||$ as the *norm* of M. For example,

and
$$||A|| = 5 \quad \text{if } A = \begin{bmatrix} -1 & 2 \\ 1 & 1 \end{bmatrix}$$

$$||x|| = 3 \quad \text{if } x = \begin{bmatrix} -1 \\ 0 \\ 2 \end{bmatrix}$$

Exercises

5. Show that for all $k \times \ell$ matrices A and B:

a. $||A+B|| \leq ||A|| + ||B||$ (the triangle inequality)

b. $||\sigma A|| = |\sigma| \, ||A||$ for all scalars σ (homogeneity)

c. $||A|| = 0$ iff $A = 0$

d. If A is $k \times \ell$ and B is $\ell \times r$, then $||AB|| \leq ||A|| \, ||B||$

6. Show that $|A|$ (the spectral radius of the $k \times k$ matrix A) doesn't satisfy all four of conditions (a) through (d)

above by finding counterexamples for the one(s) which it
doesn't satisfy.

7. a. If Z_1, Z_2, ..., Z_n, ... is a convergent sequence of
        $k \times \ell$ matrices, prove that $\lim\limits_{n\to\infty}||Z_n|| = ||\lim\limits_{n\to\infty} Z_n||$.

    b. Show that $\lim\limits_{n\to\infty} A_n = A$ iff $\lim\limits_{n\to\infty}||A_n-A|| = 0$.

8. Show that $\left|\left|\int_\alpha^\beta M(s)\,ds\right|\right| \le \int_\alpha^\beta ||M(s)||\,ds$ iff $\beta \ge \alpha$.

We say that a sequence of $k \times \ell$ matrices A_1, A_2, ...,
A_n, ... is a *Cauchy* sequence iff for every $\varepsilon > 0$ there is an
integer $\nu(\varepsilon)$ such that $||A_n - A_m|| < \varepsilon$, whenever $n \ge \nu(\varepsilon)$
and $m \ge \nu(\varepsilon)$. You may recall from your advanced calculus
course that a sequence of numbers converges iff it is a
Cauchy sequence. [See for example, Fulks (1961, pp. 117-118,
189-191) and Knopp (1952, pp. 73, 77).]

Exercise

9. a. Show that A_1, A_2, ..., A_n, ... is a Cauchy sequence
        iff for each i and j: $a_{ij}^{(1)}$, $a_{ij}^{(2)}$, ..., $a_{ij}^{(n)}$, ... is
        a Cauchy sequence.

    b. Prove that A_1, A_2, ..., A_n, ... converges iff it is
        a Cauchy sequence.

3. LINEAR SCALAR DIFFERENTIAL EQUATIONS

Given scalar-valued functions $a_{ij}(t)$ ($1 \le i, j \le k$) and
scalars c_1, c_2, ..., c_k, a frequently recurring problem in
applied mathematics is to find and/ or describe the solution

functions $x_i(t)$ simultaneously satisfying the k linear first
order differential equations:

$$\frac{dx_i(t)}{dt} = \sum_{j=1}^{k} a_{ij}(t)x_j(t) + g_i(t) \quad (i = 1, 2, \ldots, k) \quad (1)$$

subject to the initial conditions:

$$x_i(0) = c_i \quad (i = 1, 2, \ldots, k) \tag{2}$$

We can write these equations as

$$\frac{dx}{dt} = A(t)x(t) + g(t) \tag{1}$$

$$x(0) = c \tag{2}$$

if we put $A(t) = [a_{ij}(t)]$, $x(t) = [x_1(t), \ldots, x_k(t)]^{tr}$,
$c = [c_1, c_2, \ldots, c_k]^{tr}$, and $g(t) = [g_1(t), g_2(t), \ldots, g_k(t)]^{tr}$.
For example, the equations

$$\frac{dx_1}{dt} = tx_1 + x_2$$

$$\frac{dx_2}{dt} = 2x_1 + tx_2 + t \tag{1}$$

subject to

$$x_1(0) = 1 \quad x_2(0) = -1 \tag{2}$$

can be written as

$$\frac{dx}{dt} = \begin{bmatrix} t & 1 \\ 2 & t \end{bmatrix} x + \begin{bmatrix} 0 \\ t \end{bmatrix} \tag{1}$$

$$x(0) = [1, -1]^{tr} \tag{2}$$

The first order matrix-vector differential equation $\frac{dx}{dt} =$ Ax + g(t) can also describe the nth order linear scalar differential equation (where y and f are scalar-valued):

$$\frac{d^n y}{dt^n} + a_n(t) \frac{d^{n-1} y}{dt^{n-1}} + \cdots + a_2(t) \frac{dy}{dt} + a_1(t)y = f(t)$$

with initial conditions

$$\left(\frac{d^j y}{dt^j}\right)_{t=0} = c_j \quad (i = 0, 1, \ldots, n-1)$$

because we can write them as:

$$\frac{d}{dt}\begin{bmatrix} x_1(t) \\ x_2(t) \\ \vdots \\ x_n(t) \end{bmatrix} = \begin{bmatrix} 0 & 1 & 0 & 0 & \cdots & 0 \\ 0 & 0 & 1 & 0 & \cdots & 0 \\ \cdot & \cdot & \cdot & \cdot & \cdots & \cdot \\ 0 & 0 & 0 & 0 & \cdots & 1 \\ -a_1(t) & -a_2(t) & -a_3(t) & -a_4(t) & \cdots & -a_n(t) \end{bmatrix}\begin{bmatrix} x_1(t) \\ x_2(t) \\ \vdots \\ x_n(t) \end{bmatrix} + \begin{bmatrix} 0 \\ 0 \\ \cdot \\ 0 \\ f(t) \end{bmatrix}$$

x(0) = c

if we put $x_1(t) = y(t)$, $x_2(t) = dy/dt$, $x_3(t) = d^2 y/dt^2$, ..., $x_n(t) = d^{n-1}y/dt^{n-1}$; and $c = [c_0, c_1, \ldots, c_{n-1}]^{tr}$.

Example: The second order linear scalar differential equation

$$\frac{d^2 y}{dt^2} - t^2 \frac{dy}{dt} + e^t y = t^3$$

subject to

$$y = 1 \text{ and } \frac{dy}{dt} = -1 \text{ at } t = 0$$

becomes

$$\frac{d}{dt}\begin{bmatrix} x_1(t) \\ x_2(t) \end{bmatrix} = \begin{bmatrix} 0 & 1 \\ -e^t & t^2 \end{bmatrix}\begin{bmatrix} x_1(t) \\ x_2(t) \end{bmatrix} + \begin{bmatrix} 0 \\ t^3 \end{bmatrix}$$

$$x(0) = \begin{bmatrix} 1 \\ -1 \end{bmatrix}$$

if we put $y(t) = x_1(t)$, $dy/dt = x_2(t)$.

The material developed in the next section will enable us to prove (in Sec. 5) that such n^{th} order scalar problems always have a (unique) solution if the functions $a_j(t)$ and $f(t)$ are continuous.

4. HOMOGENEOUS DIFFERENTIAL EQUATIONS: THE FUNDAMENTAL MATRIX

We have seen that all the problems mentioned in Sec. 3 can be expressed in the form:

$$\frac{dx}{dt} = A(t)x(t) + g(t)$$

(3)

$$x(0) = c$$

where $A(t)$, $g(t)$, and c are known and $x(t)$ is to be determined. As a first step towards solving Eq. (3) we'll consider the simpler problem:

$$\frac{dx}{dt} = A(t)x(t)$$

(4)

$$x(0) = c$$

[The differential equation $dx/dt = A(t)x(t)$ is said to be *homogeneous*].

If we could find a k × k matrix function $X(t)$ satisfying

$$\frac{dX}{dt} = A(t)X(t) \qquad\qquad (4a)$$

and

$$X(0) = I \qquad\qquad (4b)$$

then x(t) = X(t)c would be a solution to Eq. (4) because we would have

$$\frac{dx}{dt} = \frac{dX}{dt}\,c \qquad \text{(by Exercise 2)}$$
$$= A(t)X(t)c \quad \text{[by (4a)]}$$
$$= A(t)x(t)$$

and

$$x(0) = X(0)c$$
$$= Ic \qquad \text{[by (4b)]}$$
$$= c$$

A matrix-valued function X(t) satisfying Eq. (4a) and (4b) is called a *fundamental matrix for the differential equation* dX/dt = AX; it is also called a *fundamental matrix for* A.

When can we be sure that there is such a convenient matrix X and how do we find it? Continuous matrix functions A(t) always have a fundamental matrix as we shall prove in the Existence lemma. A method of finding the fundamental matrix (or approximating it at any rate) is contained in the proof of the lemma. This leads to a simple formula for the fundamental matrix when A is constant (see Example 1).

Although A(t) is frequently defined for all real t, it may happen that it is defined only on some interval or ray[†] containing zero. So hereafter we'll assume that A(t) is defined on a set Ω, which can be the whole real line, an interval or a ray, and we'll also assume that $0 \ \epsilon \ \Omega$.

Existence Lemma: If A(t) is a k × k matrix-valued function, continuous at each $t \ \epsilon \ \Omega$, then there exists a fundamental matrix for A(t) defined at each $t \ \epsilon \ \Omega$; that is, there exists a matrix-valued function X(t) satisfying

a. $\frac{dX}{dt} = A(t)X(t)$ all $t \ \epsilon \ \Omega$

b. X(0) = I

Proof: Let $\mu(t) = \text{lub}\{||A(\tau)||:|\tau| \leq |t|\}$ for all $t \ \epsilon \ \Omega$, then $\mu(s) \leq \mu(t)$, when $|s| \leq |t|$. Define for all $t \ \epsilon \ \Omega$:

$$X_0(t) = I$$

$$X_1(t) = I + \int_0^t A(s)X_0(s) \ ds$$

$$X_2(t) = I + \int_0^t A(s)X_1(s) \ ds$$

$$X_3(t) = I + \int_0^t A(s)X_2(s) \ ds$$

$$\vdots$$

$$X_{n+1}(t) = I + \int_0^t A(s)X_n(s) \ ds$$

$$\vdots$$

Assume first that $t \geq 0$. Then by applying Exercise 8, we obtain

---

[†] A *ray* is a set of the form $\{t \ \epsilon \ R: t \ o \ \alpha\}$, where o is $\geq$, >, $\leq$ or <.

$$||X_1(t) - X_0(t)|| = ||\int_0^t A(s)ds||$$

$$\le \int_0^t ||A(s)||ds \le \mu(t)\int_0^t ds \le \mu(t)t \tag{5}$$

Applying Exercises 8 and 5(d) we have

$$||X_2(t) - X_1(t)|| = ||\int_0^t A(s)[X_1(s) - X_0(s)]ds||$$

$$\le \int_0^t ||A(s)||\,||X_1(s) - X_0(s)||ds$$

$$\le \int_0^t \mu(t)\mu(t)sds$$

(as $||A(s)|| \le \mu(t)$ when $0 \le s \le t$
and for all $s \ge 0$:

$$||X_1(s) - X_0(s)|| \le \mu(s)s \text{ [from Eq. (5)]}$$
$$\le \mu(t)s)$$

$$\le \mu^2(t)\int_0^t sds$$

$$\le \frac{\mu^2(t)t^2}{2} \quad \text{for all } 0 \le t \,\varepsilon\, \Omega$$

$$||X_3(t) - X_2(t)|| \le \int_0^t ||A(s)||\,||X_2(s) - X_1(s)||ds$$

$$\le \int_0^t \mu(t)\,\frac{\mu^2(t)s^2}{2}\,ds$$

$$\le \frac{\mu^3(t)t^3}{3!} \quad \text{for all } 0 \le t \,\varepsilon\, \Omega$$

By induction we obtain:

$$||X_{n+1}(t) - X_n(t)|| \le \frac{\mu^{n+1}(t)t^{n+1}}{(n+1)!}$$

for all $0 \le t \,\varepsilon\, \Omega$ and all $n \ge 0$. Similarly, if $0 > t \,\varepsilon\, \Omega$ we
can obtain $||X_{n+1}(t) - X_n(t)|| \le \frac{\mu^{n+1}(t)(-t)^{n+1}}{(n+1)!}$, therefore

$$||X_{n+1}(t) - X_n(t)|| \le \frac{\mu^{n+1}(t)|t|^{n+1}}{(n+1)!} \tag{6}$$

for all $t \,\varepsilon\, \Omega$ and all $n \ge 0$. It follows that

$$||X_{n+\ell}(t) - X_n(t)|| \leq \sum_{j=n+1}^{n+\ell} \frac{\mu^j(t)|t|^j}{j!} \qquad (7)$$

for all n, $\ell \geq 0$ and all $t \in \Omega$ because

$$||X_{n+\ell}(t) - X_n(t)|| = \left|\left|\sum_{j=n}^{n+\ell-1} X_{j+1}(t) - X_j(t)\right|\right| \leq$$

$$\sum_{j=n}^{n+\ell-1} \frac{\mu^{j+1}(t)|t|^{j+1}}{(j+1)!}$$

Therefore

$$||X_m(t) - X_n(t)|| \leq \sum_{j=n+1}^{\infty} \frac{[\mu(t)|t|]^j}{j!} \qquad (8)$$

$$\leq \frac{e^{\mu(t)|t|}|t|^{n+1}\mu(t)^{n+1}}{(n+1)!}$$

whenever $m \geq n$ and $t \in \Omega$ because, by Taylor's theorem

$$e^\theta = \sum_{j=0}^{n} \frac{\theta^j}{j!} + \frac{e^{\theta'}\theta^{n+1}}{(n+1)!}$$

for some θ' between 0 and θ; therefore

$$\sum_{j=0}^{\infty} \frac{\theta^j}{j!} - \sum_{j=0}^{n} \frac{\theta^j}{j!} = \frac{e^{\theta'}\theta^{n+1}}{(n+1)!}$$

and hence

$$0 \leq \sum_{j=n+1}^{\infty} \frac{\theta^j}{j!} = \frac{e^{\theta'}\theta^{n+1}}{(n+1)!} \leq \frac{e^\theta \theta^{n+1}}{(n+1)!} \quad \text{when } \theta > 0$$

We now that $\lim_{n\to\infty} \frac{|\tau|^n}{n!} = 0$ so, given $\varepsilon > 0$ and $t \in \Omega$, there is

an integer $\nu(\varepsilon, t)$ such that $||X_m(t) - X_n(t)|| < \varepsilon$, whenever
n and m are $\geq \nu$. Therefore the sequence of matrices
$X_1(t)$, $X_2(t)$, ..., $X_n(t)$, ... is a Cauchy sequence for each
$t \in \Omega$. It must converge to some matrix. As that matrix
depends on the choice of $t \in \Omega$ we call it X(t). The function
we have just defined is the one we require, as we shall now
show.

Exercise

10. For every t and t' in Ω and every $n \geq 0$, show that

a. $||X_n(t)|| \leq ke^{\mu(t)|t|}$

b. $||X_n(t) - X_n(t')|| \leq k\mu(t)e^{\mu(t)|t|}|t - t'|$ if
$|t'| \leq |t|$, and hence that

c. X(t) is continuous at all $t \in \Omega$.

As X is continuous we may integrate it, obtaining

$$||X(t) - (I + \int_0^t A(s)X(s)\ ds)||$$

$$= ||X(t) - X_n(t) - (I + \int_0^t A(s)X(s)\ ds) + I$$

$$+ \int_0^t A(s)X_{n-1}(s)\ ds||$$

$$\leq ||X(t) - X_n(t)|| + ||\int_0^t A(s)\left[X(s) - X_{n-1}(s)\right]ds||$$

Applying Eq. (8) we have

$$||X(t) - X_n(t)|| \le \frac{e^{\mu(t)|t|}|t|^{n+1}\mu(t)^{n+1}}{(n + 1)!} \quad \text{for all } t \in \Omega \quad (9)$$

Thus

$$||X(t) - [I + \int_0^t A(s)X(s) \, ds]|| \le \frac{2e^{\mu(t)|t|}|t|^{n+1}\mu(t)^{n+1}}{(n + 1)!}$$

$$\text{for all } n$$

As the right member of the inequality tends to 0 as $n \to \infty$, the left member must be zero, and hence

$$X(t) = I + \int_0^t A(s)X(s) \, ds$$

Consequently $X(0) = I$ and $dX/dt = A(t)X(t)$ for all $t \in \Omega$. []

Uniqueness Lemma: If A is as in the previous lemma,

$$\frac{dZ}{dt} = A(t)Z(t) \quad \text{for all } t \in \Omega$$

and

$$Z(0) = 0$$

then

$$Z(t) = 0 \quad \text{for all } t \in \Omega$$

Proof:

$$Z(t) = \int_0^t A(s)Z(s) \, ds \quad \text{for all } t \in \Omega$$

Let $\zeta(t) = \text{lub}\{||Z(\tau)|| : |\tau| \le |t|\}$, then $\zeta(s) \le \zeta(t)$ when $|s| \le |t|$. Assume first that $0 \le t \in \Omega$

$$||Z(t)|| \leq ||\int_0^t A(s)Z(s)\ ds||$$

$$\leq \int_0^t \mu(t)\zeta(t)\ ds \leq \mu(t)\zeta(t)t \quad \text{for all } 0 \leq t \ \epsilon \ \Omega$$

$$||Z(t)|| \leq ||\int_0^t A(s)Z(s)\ ds||$$

$$\leq \int_0^t \mu(t)\mu(t)\zeta(t)s\ ds$$

as $||Z(s)|| \leq \mu(s)\zeta(s)s$ and $||A(s)|| \leq \mu(t)$. Therefore

$$||Z(t)|| \leq \frac{\mu^2(t)\zeta(t)t^2}{2} \quad \text{for all } 0 \leq t \ \epsilon \ \Omega$$

$$||Z(t)|| \leq ||\int_0^t A(s)Z(s)\ ds||$$

$$\leq \int_0^t \mu(t)\mu^2(t)\zeta(t)\frac{s^2}{2}\ ds$$

as $||Z(s)|| \leq \dfrac{\mu^2(s)\zeta(s)s^2}{2} \leq \dfrac{\mu^2(t)\zeta(t)s^2}{2}$. Therefore

$$||Z(t)|| \leq \frac{\mu^3(t)\zeta(t)t^3}{3!}$$

$$\vdots$$

$$||Z(t)|| \leq \frac{\mu^n(t)\zeta(t)t^n}{n!} \quad \text{for all } n \text{ when } 0 \leq t \ \epsilon \ \Omega$$

A similar argument for negative t in Ω can be employed to obtain

$$||Z(t)|| \leq \frac{\mu^n(t)\zeta(t)|t|^n}{n!}$$

for all n and all $t \ \epsilon \ \Omega$. Therefore $Z(t) = 0$ for all $t \ \epsilon \ \Omega$.[]

Let us summarize the results obtained so far:

Theorem 1. If A(t) is a continuous k × k matrix-valued
function on Ω (an interval or ray of $\underline{R}$ or $\underline{R}$ itself) and $0 \in \Omega$,
then there exists one, and only one, k × k matrix-valued
function X such that

a. $\frac{dX}{dt} = A(t)X(t)$ for all $t \in \Omega$

b. $X(0) = I$

(In short, A has a unique fundamental matrix on Ω if A is
continuous on Ω).

Proof: The Existence lemma supplies us with one function
X which will do. If $dY/dt = A(t)Y(t)$ for all $t \in \Omega$ and
$Y(0) = I$, then define $Z(t) = X(t) - Y(t)$ for all $t \in \Omega$. Z
satisfies the hypotheses of the Uniqueness lemma. Therefore,
$Z(t) = 0 = X(t) - Y(t)$ for all $t \in \Omega$, and hence $Y = X$.

Corollary. Under the hypotheses of theorem 1, given any
k × 1 vector c, there exists a unique vector-valued function
x(t) such that

$$\frac{dx}{dt} = A(t)x(t) \quad \text{for all } t \in \Omega$$

$$x(0) = c$$

Proof: For existence, use $x(t) = X(t)$ c -- where X(t)
is the fundamental matrix for A(t). For uniqueness: if y(t)
is another solution, let $z(t) = x(t) - y(t)$ and Z(t) be the
square matrix each of whose columns is z. It follows that
$dZ/dt = AZ$ and $Z(0) = 0$. Therefore, $Z(t) = 0$ for all $t \in \Omega$,
and hence $z(t) = 0$ for all $t \in \Omega$. Consequently, $y = x$ and
there is only one solution.

Example 1: Suppose A is constant, then

$$X_0 = I$$

$$X_1 = I + \int_0^t A \, ds = I + At$$

$$X_2 = I + \int_0^t A(As + I) \, ds = I + At + \frac{A^2 t^2}{2}$$

$$\vdots$$

$$X_n = I + At + \frac{A^2 t^2}{2} + \cdots + \frac{A^n t^n}{n!}$$

$$\vdots$$

Therefore $X = e^{At}$ is the fundamental matrix for A when A is constant.

Example 2: Suppose $A = \begin{bmatrix} t & 1 \\ 2 & t \end{bmatrix}$ for all t. Given $dx/dt = Ax$ and $x(0) = [1, -1]^{tr}$, we'll estimate x(t) and analyze the error for $|t| \le 0.1$.

$$X_0 = I$$

$$X_1 = I + \int_0^t \begin{bmatrix} s & 1 \\ 2 & s \end{bmatrix} \, ds = I + \begin{bmatrix} \dfrac{t^2}{2} & t \\ 2t & \dfrac{t^2}{2} \end{bmatrix}$$

$$X_2 = I + \int_0^t \begin{bmatrix} \dfrac{s^3}{2} + 3s & \dfrac{3s^2}{2} + 1 \\ 3s^2 + 2 & \dfrac{s^3}{2} + 3s \end{bmatrix} \, ds =$$

$$\begin{bmatrix} \dfrac{t^4}{8} + \dfrac{3t^2}{2} + 1 & \dfrac{t^3}{2} + t \\ t^3 + 2t & \dfrac{t^4}{8} + \dfrac{3t^2}{2} + 1 \end{bmatrix}$$

$$||X(t) - X_2(t)|| \le \frac{e^{\mu(t)|t|} |t|^3 \mu^3(t)}{3!}$$ by Eq. (9) of the proof of the Existence lemma. If we approximate the solution

$x = Xc$ $(c = [1, -1]^{tr})$ by X_2c, then

$$||x(t) - X_2(t)c|| \leq ||X(t) - X_2(t)||\,||c|| \quad \text{[by Exercise}$$
$$5(d)]$$

$$\leq \frac{2}{3!}\, e^{\mu(t)|t|}|t|^3\mu^3(t)$$

But $(0.1)\mu(0.1) = 0.32$ as

$$\mu(0.1) = \sup\{||A(t)|| : |t| \leq 0.1\} =$$

$$\sup\{2|t| + 3 : |t| \leq 0.1\} = 3.2$$

When $|t| \leq 0.1$ we have

$$||x(t) - X_2(t)c|| \leq (\tfrac{1}{3})(e^{0.32})(0.32)^3 < \tfrac{1}{3}(2)(\tfrac{1}{3})^3$$

$$= \frac{2}{81} < 0.025$$

and

$$X_2(t)c = \begin{bmatrix} \dfrac{t^4}{8} - \dfrac{t^3}{2} + \dfrac{3t^2}{2} - t + 1 \\[3mm] -\dfrac{t^4}{8} + t^3 - \dfrac{3t^2}{2} + 2t - 1 \end{bmatrix}$$

Therefore for all $|t| \leq 1$,

$$-0.026 < -0.025 - \frac{(0.1)^3}{2} < x_1(t) - (1 - t + \frac{3t^2}{2}) <$$

$$< 0.025 - \frac{(-0.1)^3}{2} + \frac{(0.1)^4}{4} < 0.026$$

$$-0.027 < -0.025 - \frac{(0.1)^4}{8} - (0.1)^3 < x_2(t) - (-1 + 2t - \frac{3t^2}{2}) <$$

$$< 0.025 + (0.1)^3 < 0.026$$

Consequently,

$$x(t) = \begin{bmatrix} 1 - t + \frac{3t^2}{2} \pm 0.026 \\ \\ -1 + 2t - \frac{3t^2}{2} \pm 0.027 \end{bmatrix} \quad \text{for all } |t| \le 0.1$$

Exercises

11. If

$$\frac{dx}{dt} = \begin{bmatrix} t & 1 \\ 0 & t \end{bmatrix} x$$

$$x(0) = \begin{bmatrix} .1 \\ .7 \end{bmatrix}$$

for all t, find x(0.1) to two decimal places.

12. Find the fundamental matrix for $dX/dt = \begin{bmatrix} 0 & 1 & t \\ 0 & 0 & 1 \\ 0 & 0 & 0 \end{bmatrix} X$.

13. Find dy/dt at t = 0.9 (to two decimal places) if

$$\frac{d^3y}{dt^3} - t^2\frac{d^2y}{dt^2} - ty = 0$$

given y = 0.1, dy/dt = 0.1, and d^2y/dt^2 = 0 when t = 1.

Explain why your estimate is correct to two decimal places.

14. Suppose A(s)A(t) = A(t)A(s) and A(t) is continuous for all s, t ε Ω. Prove that $\exp[\int_0^t A(s) ds]$ is the

fundamental matrix of A(t). (Suggestion: Let B(t) = $\int_0^t A(s) ds$ for all t ε Ω, and prove the following sequence

of propositions:

a. B(s)B(t) = B(t)B(s) for all s, t ε Ω

b. $\frac{dB}{dt} B = B \frac{dB}{dt}$ at every t ε Ω

c. $\frac{d}{dt}(B^n) = n(\frac{dB}{dt}) B^{n-1}$

   for all integers $n \geq 0$ at every $t \in \Omega$

d. $\frac{d}{dt}(\sum_{m=0}^{n} \frac{B^m}{m!}) = (\frac{dB}{dt}) \sum_{m=0}^{n-1} \frac{B^m}{m!}$ at every $t \in \Omega$

Then, letting $S_n = \sum_{m=0}^{n} B^m/m!$, $M' = \frac{dM}{dt}$ and $S = e^B$, show

that for all $t \in \Omega$,

e. $\lim_{n\to\infty} \int_0^t S_n(s) ds = \int_0^t S(s) ds$ (Why is $S(s)$ continuous?)

f. $\frac{dS}{dt} = \frac{dB}{dt} S$ [and hence that $e^{\int_0^t A(s) ds}$ is the

   fundamental matrix of $A(t)$].

[*Hint for (e)*: Show that $\left\| \int_0^t S_n(s) ds - \int_0^t S(s) ds \right\| \leq$

$|t| \sum_{m=n+1}^{\infty} \beta^m/m!$ if $t \in \Omega$ and $\beta = \underset{|s| \leq |t|}{\text{lub}} \|B(s)\|$]

[*Hint for (f)*: $S_n(t) - S_n(0) = \int_0^t S_n'(s) ds$ (Why?)

Therefore $S(t) - S(0) = \lim_{n\to\infty} \int_0^t S_n'(s) ds$ (Why?)

$= \lim_{n\to\infty} \int_0^t B'(s)S_{n-1}(s) ds$ [by (d)]

$= \int_0^t B'(s)S(s) ds$ (Why?)]

Example 3. Suppose, as in Exercise 2, that $A = \begin{bmatrix} t & 1 \\ 2 & t \end{bmatrix}$

and we want to solve

$$\frac{dx}{dt} = Ax$$
$$x(0) = [1, -1]^{tr}$$

If we let $C = \begin{bmatrix} 0 & 1 \\ 2 & 0 \end{bmatrix}$, then $A(t) = tI + C$, so it is easy to
see why $A(s)A(t) = A(t)A(s)$ for all s, t, and hence

(Exercise 14) the fundamental matrix of A is

$$X = \exp[\int_0^t A(s)ds]$$

$$= e^{(t^2/2)I+tC}$$

$$= e^{(t^2/2)}e^{tC} \quad \text{(by Exercise 67 of Chap. 1)}$$

But

$$e^{tC} = \begin{bmatrix} \cosh t\sqrt{2} & \frac{1}{\sqrt{2}}\sinh t\sqrt{2} \\ \sqrt{2}\sinh t\sqrt{2} & \cosh t\sqrt{2} \end{bmatrix}$$

as you may determine by the method of Chap. 1, Sec. 15.   Here
is another way to see it:

$$e^{tC} = \sum_{n=0}^{\infty} \frac{t^n C^n}{n!}$$

$$= \sum_{m=0}^{\infty} \frac{t^{2m}C^{2m}}{(2m)!} + \frac{t^{2m+1}C^{2m+1}}{(2m+1)!}$$

$$= \sum_{m=0}^{\infty} \frac{t^{2m}2^m}{(2m)!} I + \sum_{m=0}^{\infty} \frac{t^{2m+1}2^m}{(2m+1)!} C \quad \text{as } C^{2m} = 2^m I$$

$$= (\cosh t\sqrt{2})I + (1/\sqrt{2})(\sinh t\sqrt{2})C$$

Therefore

$$x = X[1,-1]^{tr}$$

$$= e^{t^2/2} \begin{bmatrix} \cosh t\sqrt{2} & -\frac{1}{\sqrt{2}}\sinh t\sqrt{2} \\ \sqrt{2}\sinh t\sqrt{2} & -\cosh t\sqrt{2} \end{bmatrix} \quad \text{for all } t \; \varepsilon \; \Omega$$

is the "closed-form" solution.   Compare this with Exercise
2:   If we expand $e^{t^2/2}$, $\sinh t\sqrt{2}$, and $\cosh t\sqrt{2}$ in their
Maclaurin series we obtain

$$x = e^{t^2/2} \begin{bmatrix} 1 + t^2 + \dfrac{t^4}{6} + \cdots & -\dfrac{1}{\sqrt{2}}\left(t\sqrt{2} + \dfrac{t^3\sqrt{2}}{2} + \cdots\right) \\ \sqrt{2}\left(t\sqrt{2} + \dfrac{t^3\sqrt{2}}{3} + \cdots\right) & -\left(1 + t^2 + \dfrac{t^4}{6} + \cdots\right) \end{bmatrix}$$

$$= \left(1 + \dfrac{t^2}{2} + \cdots\right) \begin{bmatrix} 1 - t + t^2 - \dfrac{t^3}{3} + \dfrac{t^4}{6} - \cdots \\ -1 + 2t - t^2 + \dfrac{2t^3}{3} - \dfrac{t^4}{6} + \cdots \end{bmatrix}$$

$$= \begin{bmatrix} 1 - t + \dfrac{3t^2}{2} - \dfrac{5t^3}{6} + \cdots \\ -1 + 2t - \dfrac{3t^2}{2} + \dfrac{3t^3}{2} - \cdots \end{bmatrix}$$

Exercises

15. a. Use the method of Example 3 to find a "closed-form" solution to the problem of Exercise 11.

   b. Can the method of Example 3 be used to solve the problem of Exercise 13?

   c. Explain why the method of Example 3 cannot be used to find the fundamental matrix of $\begin{bmatrix} 1 & t \\ 2 & t \end{bmatrix}$.

   d. For each of the following matrices, find the fundamental matrix if the method of Example 3 can be employed to do so. If the method can't be used, explain why it can't.

$$\begin{bmatrix} 1 & t \\ t & 1 \end{bmatrix} \quad \begin{bmatrix} 1 & t^2 \\ t & 1 \end{bmatrix} \quad \begin{bmatrix} 1 & -t \\ t & 1 \end{bmatrix} \quad \begin{bmatrix} 1 & t & 0 \\ 0 & 1 & t \\ t & 0 & 1 \end{bmatrix}$$

Jacobi's Identity

   We have seen that when $A(t)$ is continuous on Ω, the differential equation $dX/dt = A(t)X(t)$ $(t \in \Omega)$ has exactly one solution X satisfying $X(0) = I$. This matrix (the fundamental matrix of A) has the interesting property that

$$\det X(t) = \exp(\int_0^t \text{trace}[A(s)] \, ds) \quad \text{for all } t \; \varepsilon \; \Omega \quad (10)$$

Equation (10) is known as *Jacobi's Identity*. In order to prove it we need the following:

   Lemma. If M(t) is differentiable at t and M_i is the ith row of M(t), then

$$\frac{d}{dt} [\det M(t)] = \sum_{i=1}^{k} \det \begin{bmatrix} M_1 \\ \vdots \\ M_{i-1} \\ \dfrac{dM_i}{dt} \\ M_{i+1} \\ \vdots \\ M_k \end{bmatrix}$$

   Proof: You may recall that $\det(M) = \sum_{\sigma} sg(\sigma) \prod_{j=1}^{k} m_{j\sigma(j)}$ where the summation is taken over all permutations σ of $\{1, 2, \ldots, k\}$, and $sg(\sigma)$ is 1 if σ is an even permutation and $sg(\sigma) = -1$ if σ is an odd permutation [see e.g., Lipschutz (1968, pp. 171-172).]

   Letting y' denote dy/dt, we have

$$(\det M)' = \sum_{\sigma} sg(\sigma)(\prod_{j=1}^{k} m_{j\sigma(j)})'$$

$$= \sum_{\sigma} sg(\sigma) \sum_{i=1}^{k} \left[\prod_{j=1}^{i-1} m_{j\sigma(j)} \right] m'_{i\sigma(i)} \left[\prod_{j=i+1}^{k} m_{j\sigma(j)} \right]$$

$$= \sum_{i=1}^{k} \sum_{\sigma} sg(\sigma) m_{1\sigma(1)} \cdots m_{i-1,\sigma(1-1)} m'_{i\sigma(i)} m_{i+1,\sigma(i+1)}$$

$$\cdots m_{k\sigma(k)}$$

$$= \sum_{i=1}^{k} \det \begin{bmatrix} M_1 \\ \vdots \\ M_{i-1} \\ M_i' \\ M_{i+1} \\ \vdots \\ M_k \end{bmatrix}$$

[]

Exercise

16. Show that

$$\det \begin{bmatrix} M_1 \\ M_2 \\ \vdots \\ M_{i-1} \\ M_i \\ M_{i+1} \\ \vdots \\ M_k \end{bmatrix} + \det \begin{bmatrix} M_1 \\ M_2 \\ \vdots \\ M_{i-1} \\ N_i \\ M_{i+1} \\ \vdots \\ M_k \end{bmatrix} = \det \begin{bmatrix} M_1 \\ M_2 \\ \vdots \\ M_{i-1} \\ M_i + N_i \\ M_{i+1} \\ \vdots \\ M_k \end{bmatrix}$$

where M_i is the ith row of $k \times k$ matrix M and N_i is any $1 \times k$ row vector.

Proof of Jacobi's Identity: Applying the lemma to X(t) and using the fact that dX/dt = AX, we have:

$$\frac{d}{dt}[\det X(t)] = \sum_{i=1}^{k} \det \begin{bmatrix} X_1 \\ X_2 \\ \vdots \\ X_{i-1} \\ (AX)_i \\ X_{i+1} \\ \vdots \\ X_k \end{bmatrix}$$

But

$$(AX)_i = \left[\sum_{\ell=1}^{k} a_{i\ell} x_{\ell 1}, \ldots, \sum_{\ell=1}^{k} a_{i\ell} x_{\ell j}, \ldots, \sum_{\ell=1}^{k} a_{i\ell} x_{\ell k} \right] = \sum_{\ell=1}^{k} a_{i\ell} X_\ell$$

so by Exercise 16

$$\frac{d}{dt} [\det X(t)] = \sum_{i=1}^{k} \det \begin{bmatrix} X_1 \\ \vdots \\ X_{i-1} \\ \sum_{\ell=1}^{k} a_{i\ell} X_\ell \\ \vdots \\ X_k \end{bmatrix} = \sum_{i=1}^{k} \sum_{\ell=1}^{k} \det \begin{bmatrix} X_1 \\ \vdots \\ X_{i-1} \\ a_{i\ell} X_\ell \\ \vdots \\ X_k \end{bmatrix}$$

$$= \sum_{i=1}^{k} \sum_{\ell=1}^{k} a_{i\ell} \det \begin{bmatrix} X_1 \\ \vdots \\ X_{i-1} \\ X_\ell \\ \vdots \\ X_k \end{bmatrix}$$

When $\ell \neq i$ the matrix $\begin{bmatrix} X_1 \\ \vdots \\ X_{i-1} \\ X_\ell \\ X_{i+1} \\ \vdots \\ X_k \end{bmatrix}$ has different rows equal to the

same row vector; therefore the determinant of each of those

matrices is zero, and we have

$$\frac{d}{dt}[\det X(t)] = \sum_{i=1}^{k} a_{ii} \det \begin{bmatrix} X_1 \\ X_2 \\ \vdots \\ X_{i-1} \\ X_i \\ X_{1+1} \\ \vdots \\ X_k \end{bmatrix} = \sum_{i=1}^{k} a_{ii} \det(X)$$

$$= \det X(t) \ [\text{trace } A(t)] \quad \text{for all } t \ \epsilon \ \Omega$$

As the solution of the scalar equations

$$\frac{du}{dt} = u(t)v(t)$$
$$u(0) = 1$$

is

$$u(t) = e^{\int_0^t v(s) \, ds}$$

we have

$$\det X(t) = e^{\int_0^t \text{trace } A(s) \, ds}$$

completing the proof of Jacobi's identity. []

The scalar exponential e^τ is never zero, therefore:

Corollary. If $A(t)$ is a matrix-valued continuous function for all $t \ \epsilon \ \Omega$, then the fundamental matrix for $dX/dt = AX$ is nonsingular at each $t \ \epsilon \ \Omega$.

5.   INHOMOGENEOUS DIFFERENTIAL EQUATIONS

Suppose, in addition to the continuous matrix-valued function

A(t) and constant vector c, we are given the continuous
vector-valued function b(t) and are asked to solve

$$\frac{dx}{dt} = Ax + b$$
$$x(0) = c$$

(11)

We may use the familiar scalar technique of "variation of
parameters." First, find the fundamental matrix of dX/dt =
$A(t)X(t)$. Next we try to find a function $y(t)$ so that
$X(t)y(t)$ is a solution to Eq. (11). (You recall that if
$b(t) \equiv 0$, then Xc is the solution to Eq. (11); hence the
name "variation of parameters.") Let

$x(t) = X(t)y(t)$ (y is as yet undetermined)

$$\frac{dx}{dt} = \frac{dX}{dt} y + X \frac{dy}{dt}$$

$$= AXy + X \frac{dy}{dt}$$

$$= Ax + X \frac{dy}{dt}$$

We would like $dx/dt = Ax + b$, so we try to determine y so
that $X \frac{dy}{dt} = b(t)$, but the corollary to Jacobi's identity
ensures that X is invertible, so we need only solve dy/dt =
$X^{-1}(t)b(t)$, subject to $y(0) = c$, obtaining $y(t)$ =
$c + \int_0^t X^{-1}(s)b(s) \, ds$ and our solution to Eq. (11) is

$$x(t) = X(t)[c + \int_0^t X^{-1}(s)b(s) \, ds]$$

(12)

Exercise

17. a. Prove that Eq. (11) has only one solution.

b. Solve:

1. $dx/dt = \begin{bmatrix} 3 & -2 \\ 4 & -3 \end{bmatrix} x + \begin{bmatrix} 1 \\ 1 \end{bmatrix}$ given $x(0) = \begin{bmatrix} 0 \\ 0 \end{bmatrix}$ (use the

    result of Example 1 to find the fundamental matrix).

2. $dx/dt = \begin{bmatrix} 1 & -t \\ t & 1 \end{bmatrix} x + te^t \begin{bmatrix} 1 \\ -1 \end{bmatrix}$ given $x(0) = \begin{bmatrix} 0 \\ 0 \end{bmatrix}$ (use

    the method of Example 3 to find the fundamental matrix).

As a result of Eq. (12), Exercise 17(a) and the discussion

in Sec. 3, we see that the nth order *scalar* linear differential

equation:

$$\frac{d^n y}{dt^n} + a_n(t) \frac{d^{n-1} y}{dt^{n-1}} + \cdots + a_1(t)y = f(t) \quad (t \ \epsilon \ \Omega)$$

subject to the initial conditions

$$\left(\frac{d^j y}{dt^j} \right)_{t=0} = c_j \quad (0 \le j \le n - 1)$$

has one, and only one, solution provided $f(t)$ and the $a_i(t)$

are continuous on Ω.

6. THE MATRIX FUNCTION e^{At}: THE LAPLACE TRANSFORM

We saw, in Example 1 of Sec. 4, that when A is constant the

function e^{At} is the fundamental matrix of $dX/dt = At$. We

have also shown that $\det(e^{At}) = e^{(trace \ A)t}$, and hence e^{At}

is nonsingular for all t (see Chap. 1, Sec. 14 as well as

Chap. 2, Sec. 5). In Chap. 1, Sec. 15 we discussed how

various identities for scalar functions can be extended to

matrix functions. The scalar identity $e^{\alpha+\beta} = e^\alpha e^\beta$ for all

α, β is one which doesn't extend completely, we can however, show that $e^A e^B = e^{A+B}$ whenever $AB = BA$: $e^{(A+B)t}$ is the fundamental matrix of $dX/dt = (A + B)X$ because of our result in Example 1 of Sec. 4. On the other hand, $d/dt(e^{At}e^{Bt}) = e^{At}(Be^{Bt}) + Ae^{At}e^{Bt}$ by the same example and Exercise 2(b). $e^{At}B = Be^{At}$ because $(At)B = B(At)$ and $e^{At} = \sum\limits_{j=0}^{\infty}(At)^j/j!$. Therefore

$$\frac{d}{dt}(e^{At}e^{Bt}) = (A + B)e^{At}e^{Bt}$$

and

$$e^{Ao}e^{Bo} = I$$

but theorem 1 ensures that there is only one fundamental matrix, therefore $e^{(A+B)t} = e^{At}e^{Bt}$ for all t including t = 1. Consequently, $e^{A+B} = e^A e^B$.

Exercise

18. Find A, B such that $e^A e^B \neq e^{A+B}$.

Evaluation of e^{At}

As solutions to the differential equation $dx/dt = Ax + b$ depend on e^{At} when A is constant, it is useful to have ways of examining the actual entries in e^{At}. In Chap. 1, Sec. 16, we discussed two methods of evaluating matrix functions, in particular of evaluating e^A. The first method was to use the Lagrange-Sylvester theorem directly:

Example 1: If

$$A = \begin{bmatrix} 1 & 4 \\ 3 & 2 \end{bmatrix}$$

then

$$A = \begin{bmatrix} 1 & -4 \\ 1 & 3 \end{bmatrix} \begin{bmatrix} 5 & 0 \\ 0 & -2 \end{bmatrix} \begin{bmatrix} 1 & -4 \\ 1 & 3 \end{bmatrix}^{-1}$$

and hence by the Lagrange-Sylvester formula,

$$e^{At} = \frac{1}{7} \begin{bmatrix} 1 & -4 \\ 1 & 3 \end{bmatrix} \begin{bmatrix} e^{5t} & 0 \\ 0 & e^{-2t} \end{bmatrix} \begin{bmatrix} 3 & 4 \\ -1 & 1 \end{bmatrix}$$

$$= \frac{1}{7} \begin{bmatrix} 3e^{5t} + 4e^{-2t} & 4e^{5t} - 4e^{-2t} \\ 3e^{5t} - 3e^{-2t} & 4e^{5t} + 3e^{-2t} \end{bmatrix}$$

But we had to find the Jordan form of A and $P = \begin{bmatrix} 1 & -4 \\ 1 & 3 \end{bmatrix}$ first,

which was fairly easy since A was only a 2 × 2 matrix. In

general, suppose the Jordan form of A is $\overset{m}{\underset{i=1}{\oplus}} J_{n_i}(\lambda_i)$. Fix t.

Letting $f(\tau) = e^{t\tau}$ we have, by the Lagrange-Sylvester formula,

$$f(J_n(\lambda)) = e^{t\lambda}I + te^{t\lambda}U + \frac{t^2 e^{t\lambda}}{2!} U^2 + \cdots + \frac{t^n e^{t\lambda}}{n!} U^n$$

$$= e^t \left(I + tU + \frac{(tU)^2}{2} + \cdots + \frac{(tU)^{n-1}}{(n-1)!} \right)$$

Therefore

$$f(A) = e^{At} = P \left[\overset{m}{\underset{i=1}{\oplus}} e^{t\lambda_i} \sum_{j=0}^{n_i-1} \frac{(tU_{n_i})^j}{j!} \right] P^{-1} \text{ for some P} \qquad (13)$$

As we observed in Chap. 1, Sec. 16 this formula has
several drawbacks, one of which is the necessity of finding
P. We presented an indirect method of applying the Lagrange-
Sylvester theorem to·avoid this. We would now like to

present a third method. We first observe that, as a consequence
of Eq. (13),there are constant k × k matrices $A_{i\ell}$ such that

$$e^{At} = \sum_{i=1}^{m} \sum_{\ell=0}^{n_i-1} \frac{t^\ell}{\ell!} e^{\lambda_i t} A_{i\ell} \tag{14}$$

(The A_i are the matrices $P\left(\bigoplus_{j=1}^{m} M_{ij}\right)P^{-1}$ where $M_{ij} = O_{n_j}$ if

$j \neq 1$ and $M_{ii} = U_{n_i}^\ell$. For example, if

$$A = P[J_3(2) \oplus J_2(3)]P^{-1}$$

then

$$\begin{aligned}
e^{At} &= e^{2t}P(I_3 \oplus O_2)P^{-1} + te^{2t}P(U_3 \oplus O_2)P^{-1} \\
&\quad + \frac{t^2}{2} e^{2t}P(U_3^2 \oplus O_2)P^{-1} + e^{3t}P(O_3 \oplus I_2)P^{-1} \\
&\quad + te^{3t}P(O_3 \oplus U_2)P^{-1} \\
&= e^{2t}A_{10} + te^{2t}A_{11} + \frac{t^2}{2} e^{2t}A_{12} + e^{3t}A_{20} + te^{3t}A_{21}
\end{aligned}$$

Next we see that Eq. (14) can be written as

$$e^{At} = \sum_{i=1}^{\nu} e^{\lambda_i t} \left(\sum_{\ell=0}^{k-1} \frac{t^\ell}{\ell!} B_{i\ell} \right) \tag{15}$$

where the $B_{i\ell}$ are constant matrices and $\lambda_1, \lambda_2, \ldots, \lambda_\nu$ is a
complete list of the ν *distinct* eigenvalues of A. Thus in
the previous example

$$\begin{aligned}
e^{At} &= e^{2t}(A_{10} + tA_{11} + \frac{t^2}{2} A_{12} + t^3 0 + t^4 0) \\
&\quad + e^{3t}(A_{20} + tA_{21} + t^2 0 + t^3 0 + t^4 0)
\end{aligned}$$

We shall use a device for evaluating the polynomials

$$p_i(t) = \left(\sum_{\ell=0}^{k-1} \frac{t^\ell}{\ell!} B_{i\ell} \right)$$

without finding the Jordan form of A first. The method
employs the *Laplace transform* which we shall now introduce:

The Laplace transform

If[†] $\lim\limits_{\tau \to \infty} \int_0^\tau M(t)\, dt$ exists (and is finite) we denote the

limit by $\int_0^\infty M(t)\, dt$. Thus $\int_0^\infty M(t)\, dt = \left[\int_0^\infty m_{ij}(t)\, dt \right]$. In

particular, if s is any complex number and X(t) is a k × k
matrix-valued function, then we define the *Laplace transform
of X* by the function L whose value at s is

$$L(s) = \int_0^\infty X(t) e^{-st}\, dt$$

whenever the integral exists. The function L(s) is often
denoted by $\mathcal{L}(X(t))$. Notice that $\mathcal{L}(X(t)) = \left[\mathcal{L}(x_{ij}(t)) \right]$ for
all s for which all the $x_{ij}(t)$ have transforms.

Lemma: If A is constant and Re(s) > Re(λ) for all
eigenvalues λ of A then $\mathcal{L}(e^{At}) = (sI - A)^{-1}$.

Proof.

$$\mathcal{L}(e^{At}) = \lim_{\tau \to \infty} \int_0^\tau e^{At} e^{-st}\, dt$$

$$= \lim_{\tau \to \infty} \int_0^\tau e^{(A-sI)t}\, dt \quad \text{(Chap. 1, Exercise 67)}$$

[†] We assume t is a real variable.

$A - sI$ is invertible and $\lim\limits_{\tau \to \infty} e^{(A-sI)\tau} = 0$ because $\text{Re}(s) >$ $\text{Re}(\lambda)$ for all eigenvalues λ of A (see Exercises 19 and 20), therefore

$$\mathscr{L}(e^{At}) = \lim_{\tau \to \infty} (A - sI)^{-1}(e^{(A-sI)\tau} - I)$$

$$= (sI - A)^{-1}$$

Exercises

19. Show that $\int_{0}^{t} e^{Bt}\, dt = B^{-1}(e^{B\tau} - I)$ when B is invertible.

20. Show that $\lim\limits_{\tau \to \infty} e^{B} = 0$ when $\text{Re}(\mu) < 0$ for every eigenvalue μ of B.

21. Show that $\mathscr{L}(t^n e^{\lambda t}) = \dfrac{n!}{(s - \lambda)^{n+1}}$ for all $n \geq 0$ and all complex s such that $\text{Re}(s) > \text{Re}(\lambda)$.

22. Assuming the Laplace transforms of the scalar function $f(t)$ and the matrix functions $X(t)$ and $Y(t)$ exist and that A is constant, show that

$$\mathscr{L}(f(t)A) = \mathscr{L}(f(t))A$$
$$\mathscr{L}(AX(t)) = A\mathscr{L}(X(t))$$
and
$$\mathscr{L}(X(t) + Y(t)) = \mathscr{L}(X(t)) + \mathscr{L}(Y(t))$$

Let us return to the problem of expressing e^{At} explicitly. Exercises 21 and 22 and Eq. (15) enable us to write

$$(e^{At}) = \sum_{i=1}^{\nu} \sum_{j=0}^{k-1} \mathscr{L}\left(\frac{t^j e^{\lambda_i t}}{j!}\right) B_{ij}$$

$$= \sum_{i=1}^{\nu} \sum_{j=0}^{k-1} \frac{1}{(s - \lambda_i)^{j+1}} B_{ij}$$

Therefore if λ_1, λ_2, ..., λ_ν are the ν distinct eigenvalues of A, then there are constant matrices B_{ij} such that:

$$(sI - A)^{-1} = \sum_{i=1}^{\nu} \sum_{j=0}^{k-1} \frac{1}{(s - \lambda_i)^{j+1}} B_{ij} \qquad (16)$$

We can use Eq. (16) to determine the constant matrices B_{ij}, and hence to express e^{At} in its "explicit" form -- the form given by Eq. (15). One way to do so would be to choose νk values for s and convert Eq. (16) into k^2 systems of νk equations in νk unknowns. Another method used is based on the technique of "partial fraction decomposition" you may recall from your first calculus course. We will also make use of the *adjugate*[†] of sI - A to evaluate $(sI - A)^{-1}$ by means of the formula:

(5) $$(sI - A)^{-1} = \frac{1}{c_A(s)} \text{adj}(sI - A) \qquad (17)$$

Example 2: If $A = \begin{bmatrix} 1 & 4 \\ 3 & 2 \end{bmatrix}$, as in Example 1 then $c_A(s) = (s - 5)(s + 2)$ so (15) implies that $e^{AT} = e^{5t}C + e^{-2t}D$ for some constant matrices $C(= B_{10})$ and $D(= B_{20})$. Applying (16) and (17) we have

---

[†] The adjugate of a k × k matrix M with k > 1, is the k × k matrix adj(M) whose ij^{th} entry is $(-1)^{i+j}$ times the determinant of the matrix obtained from M by deleting column i and row j. It has the critical property that M adj(M) = (det M)I, see e.g. Lipschutz, p.76. This property implies (17) as $c_A(s) = \det(sI - A)$.

$$\frac{1}{(s - 5)(5 + 2)} \begin{bmatrix} s - 2 & 4 \\ 3 & s - 1 \end{bmatrix} = \frac{1}{s - 5}C + \frac{1}{s + 2}D$$

so

$$\begin{bmatrix} s - 2 & 4 \\ 3 & s - 1 \end{bmatrix} = (s + 2)C + (s - 5)D$$

let s = 5 and obtain $\frac{1}{7} \begin{bmatrix} 3 & 4 \\ 3 & 4 \end{bmatrix} = C$

let s = -2 and obtain $(-\frac{1}{7}) \begin{bmatrix} -4 & 4 \\ 3 & 3 \end{bmatrix} = D$

Therefore $e^{At} = \frac{1}{7} e^{5t} \begin{bmatrix} 3 & 4 \\ 3 & 4 \end{bmatrix} - \frac{1}{7} e^{-2t} \begin{bmatrix} -4 & 4 \\ 3 & -3 \end{bmatrix}$.

More generally, if the k × k matrix A has k distinct eigenvalues then (15) implies that for some constant matrices A_i

$$e^{At} = \sum_{i=1}^{k} e^{\lambda_i t} A_i \quad \text{and hence, by (16) we have:}$$

$$(sI - A)^{-1} = \sum_{i=1}^{k} (s - \lambda_i)^{-1} A_i, \text{ but } c_A(s) = \prod_{\ell=1}^{k} (s - \lambda_\ell)$$

therefore, by (17) we have

$$\text{adj}(sI - A) = \sum_{i=1}^{k} \prod_{\ell \neq i} (s - \lambda_\ell) A_i \quad \text{and hence for each}$$

$$1 \leq j \leq k :$$

$$\text{adj}(\lambda_j I - A) = (\prod_{\ell \neq j} (\lambda_j - \lambda_\ell)) A_j . \quad \text{Consequently}$$

$$(6)\quad\begin{cases} e^{At} = \sum_{j=1}^{k} e^{\lambda_j t} \prod_{\ell \neq j} (\lambda_j - \lambda_\ell)^{-1} \text{adj}(\lambda_j I - A) \quad \text{when} \\[2mm] A \text{ is a } k \times k \text{ matrix having } k \text{ distinct eigenvalues} \\[2mm] \lambda_1, \lambda_2, \ldots, \lambda_k \end{cases}$$

Example 3: If $A = \begin{bmatrix} 0 & -1 \\ 4 & 4 \end{bmatrix}$ then $c_A(s) = (s - 2)^2$ so (15) implies that $e^{At} = e^{2t}(C + tD)$ for some constant matrices C and D, and hence by (16) and (17) we have

$$\frac{1}{(s-2)^2} \begin{bmatrix} s-4 & -1 \\ 4 & s \end{bmatrix} = \frac{1}{s-2}C + \frac{1}{(s-2)^2}D$$

Therefore

$$\begin{bmatrix} s-4 & -1 \\ 4 & s \end{bmatrix} = (s-2)C + D$$

letting $s = 2$ we get $D = \begin{bmatrix} -2 & -1 \\ 4 & 2 \end{bmatrix}$. Differentiating we get $C = I$. Therefore $e^{At} = e^{2t}I + te^{2t}\begin{bmatrix} -2 & -1 \\ 4 & 2 \end{bmatrix}$.

More generally, if $c_A(t) = (t - \lambda)^k$ then (15) implies that

$$e^{At} = \sum_{i=1}^{k} e^{\lambda t} \frac{t^i}{i!} B_i$$

and hence

$$(sI - A)^{-1} = \sum_{i=1}^{k} (s - \lambda)^{-i} B_{i-1}$$

Letting $B(s) = \text{adj}(sI - A)$ and applying (17) we have

$$B(s) = \sum_{i=1}^{k} (s - \lambda)^{k-i} B_{i-1}$$

Differentiating j times we get

$$B^{(j)}(s) = \sum_{i=1}^{k-j} (k - i)(k - i - 1) \cdots (k - i - j + 1)$$
$$(s - \lambda)^{k-i-j} B_{i-1}$$

and hence

$$B^{(j)}(\lambda) = j! \; B_{k-j-1} \text{ for } 0 \leq j \leq k-1$$

Therefore $B_i = \dfrac{1}{(k - 1 - i)} B^{(k-1-i)}(\lambda)$ for $0 \leq i \leq k - 1$.

Consequently

$$
\begin{cases}
e^{At} = e^{\lambda t} \sum_{i=0}^{k-1} (i!(k - 1 - i)!)^{-1} t^i B^{(k-1-i)}(\lambda) \\
\\
\text{when A is a } k \times k \text{ matrix having just one eigenvalue} \\
\lambda \text{ and } B^{(j)}(s) \text{ denotes the } j^{th} \text{ derivative of adj}(sI - A)
\end{cases}
$$

When $k > 2$ the matrix A may fall between the two extremes of having k distinct eigenvalues and having only one eigenvalue. Rather than presenting a formula for these intermediate cases we'll present an example to illustrate the technique of finding an explicit expression for e^{At}.

Example 4.   Suppose $c_A(s) = (s - 1)^3 (s - 3)^2 (s + 3)$.

Applying (15) we have: $e^{At} = e^t \sum_{i=0}^{2} \dfrac{t^i}{i!} C_i + e^{2t} \sum_{i=0}^{1} \dfrac{t^i}{i!} D_i + e^{-3t} E$

and hence $(sI - A)^{-1} = \dfrac{1}{(s - 1)} C_0 + \dfrac{1}{(s - 1)^2} C_1 + \dfrac{1}{(s - 1)^3} C_2 +$

$\dfrac{1}{s-3} D_0 + \dfrac{1}{(s-3)^2} D_1 + \dfrac{1}{s-3}$ E.   Letting B(s) = adj(sI - A)

we then have by (17):

$$B(s) = (s-3)^2(s+3)((s-1)^2 C_0 + (s-1)C_1 + C_2) +$$

$$(s-1)^3(s+3)((s-3)D_0 + D_1) + (s-1)^3(s-3)^2 E$$

Therefore

$$B(1) = (-2)^2(4)C_2$$

$$B(3) = 2^3(6)D_1$$

$$B(-3) = (-4)^3(-6)^2 E$$

hence $C_2 = \frac{1}{16} B(1)$, $D_1 = \frac{1}{48} B(3)$ and E $= -\frac{1}{2304} B(-3)$.
Differentiating B(s) we get

$$B'(s) = [2(s-3)(s+3) + (s-3)^2]((s-1)^2 C_0 + (s-1)C_1$$

$$+C_2) + (s-3)^2(s+3)[2(s-1)C_0 + C_1]$$

$$+[3(s-1)^2(s+3) + (s-1)^3]((s-3)D_0 + D_1)$$

$$+(s-1)^3(s+3)D_0 + [3(s-1)^2(s-3)^2$$

$$+2(s-1)^3(s-3)]E$$

Therefore B'(1) = $-12C_2 + 16C_1 = -\frac{12}{16} B(1) + 16C_1$ and

B'(3) = $80D_1 + 48D_0 = \frac{80}{48} B(3) + 48D_0$ hence $C_1 = \frac{3}{64} B(1) +$

$\frac{1}{16} B'(1)$ and $D_0 = -\frac{5}{144} B(3) + \frac{1}{48} B'(1)$. Differentiating B(s)

twice we get

$$B''(s) = [4s + 2(s-3)]((s-1)^2 C_0 + (s-1)C_1 + C_2)$$

$$+[2s^2 - 18 + (s-3)^2](2(s-1)C_0 + C_1)$$

$$+[2(s-3)(s+3) + (s-3)^2][2(s-1)C_0 + C_1]$$

$$+(s-3)^2(s+3)2C_0 + (s-1)P(s)$$

where P(s) is a polynomial in s with matrix coefficients.
Therefore

$$B''(1) = -24C_1 + 32C_0$$

$$= -24(\frac{3}{64} B(1) + \frac{1}{16} B'(1)) + 32C_0$$

hence

$$C_0 = \frac{9}{256} B(1) + \frac{3}{64} B'(1) + \frac{1}{32} B''(1)$$

Now we have found all the coefficients we needed and we have:

$$e^{At} = e^t\{(\frac{9}{256} B(1) + \frac{3}{64} B'(1) + \frac{1}{32} B''(1))$$

$$+ t(\frac{3}{64} B(1) + \frac{1}{16} B'(1)) + \frac{t^2}{32} B(1)\}$$

$$+ e^{3t}\{(-\frac{5}{144} B(3) + \frac{1}{48} B'(3)) + t(\frac{1}{48} B(3))\}$$

$$- \frac{1}{2304} B(-3)$$

Notice that the matrix coefficients C_i were all linear
combinations of B(1), B'(1), B''(1); the D_i were linear
combinations of B(3), B'(3) and E was a multiple of B(-3).
This is not an accident.

In general, if $c_A(t) = \prod_{i=1}^{\nu} (t - \lambda_i)^{m_j}$ where $\lambda_1, \lambda_2, \ldots, \lambda_\nu$
are the distinct eigenvalues of A, then (14) implies that

$$e^{At} = \sum_{i=1}^{\nu} e^{\lambda_i t} \sum_{j=0}^{m_i-1} \frac{t^j}{j!} B_{ij}$$

and it will always be the case as in our example, that each
B_{ij} is a linear combination of $B(\lambda_i)$, $B'(\lambda_i)$, $\ldots$, $B^{(m_i-1)}(\lambda_i)$.

It should now be clear that the calculation of
adj(sI - A) is a considerable part of the price paid for

avoiding the determination of a matrix which puts A into
Jordan form. Since calculating adj(sI - A) directly from its
definition can be long and arduous (even if A is only a 4 × 4
matrix), we present an alternative method.

Suppose B(s) = adj(sI - A) and A is a k × k matrix. We
know that $(sI - A)B(s) = c_A(s)I$. By repeated differentiation
we obtain

$$B(s) = (A - sI)B'(s) + c_A'(s)I$$
$$2B'(s) = (A - sI)B''(s) + c_A''(s)I$$
$$\vdots$$
$$(k - 1)B^{(k-2)}(s) = (A - sI)B^{(k-1)}(s) + c_A^{(k-1)}(s)I$$
$$kB^{(k-1)}(s) = (A - sI)B^{(k)}(s) + c_A^{(k)}(s)I$$

But every entry in B(s) is a polynomial of degree less than
k so $B^{(k)}(s) = 0$, moreover $c_A^{(k)}(s) = k!$ because $c_A(s) = s^k + a_1 s^{k-1} + \cdots + a_0$ therefore

(†)
$$
\begin{cases}
B^{(k-1)}(s) = (k - 1)!I \\[4pt]
B^{(k-2)}(s) = \left(\frac{1}{k-1}\right)((A - sI)B^{(k-1)}(s) + c_A^{(k-1)}(s)I) \\[4pt]
B^{(k-3)}(s) = (\frac{1}{k-2})((A - sI)B^{(k-2)}(s) + c_A^{(k-2)}(s)I \\[4pt]
\vdots \\[4pt]
B'(s) = \frac{1}{2}((A - sI)B''(s) + c_A''(s)I) \\[4pt]
B(s) = (A - sI)B'(s) + c_A'(s)I
\end{cases}
$$

is a system which can be solved recursively to find B(s).

Example 5: Suppose $c_A(s)$ is as in Exercise 23a below
and we want to evaluate adj(sI - A). If we use (†) to do it
then we need the first two derivatives of $c_A(s)$.

$$c_A(s) = (s - 5)(s - 1)^2$$

$$c_A'(s) = 3s^2 - 14s + 11$$

$$c_A''(s) = 6s - 14$$

Applying (†) we get

$$B''(s) = 2I$$

$$B'(s) = \frac{1}{2}((A - sI)(2I) + (6s - 14)I) = A + (2s - 7)I$$

$$B(s) = (A - sI)(A + (2s - 7)I) + (3s^2 - 14s + 11)I$$

$$= A^2 + (s - 7)A + (s^2 - 7s + 11)I$$

Exercises

23. Use the Laplace transform to express e^{At} explicitly if

a. $A = \begin{bmatrix} 2 & 2 & 1 \\ 1 & 3 & 1 \\ 1 & 2 & 2 \end{bmatrix}$ (*Note:* $c_A(t) = (t - 5)(t - 1)^2$)

b. $A = \begin{bmatrix} 1 & -4 & -1 & -4 \\ 2 & 0 & 5 & -4 \\ -1 & 1 & -2 & 3 \\ -1 & 4 & -1 & 6 \end{bmatrix}$ (*Note:* $c_A(t) = (t - 1)^3(t - 2)$)

24. Write an explicit form of the solution to both problems

a. $$\frac{dx}{dt} = \begin{bmatrix} 0 & 1 \\ -1 & 0 \end{bmatrix} x$$
$$x(0) = \begin{bmatrix} 1 \\ 3 \end{bmatrix}$$

b. $$\frac{dx}{dt} = \begin{bmatrix} 4 & 1 \\ -1 & 2 \end{bmatrix} x + e^{4t}\begin{bmatrix} 1 \\ 2 \end{bmatrix}$$
$$x(0) = \begin{bmatrix} 1 \\ -1 \end{bmatrix}$$

7. LOGARITHMS AND ARBITRARY POWERS OF A MATRIX

We have seen that e^B is a nonsingular matrix. This raises
the question: If A is a nonsingular matrix, is there a

matrix B such that e^B = A? In the case of 1 × 1 matrices you know the answer is yes, in fact there are always infinitely many (complex) solutions to e^x = a when a ≠ 0. The same is true of k × k matrices as we shall see later on.

We say that X is a *logarithm* of A if, and only if, e^X = A. It is customary to denote such matrices X by ℓn(A). Unfortunately, this labels many different objects with the same symbol which occasionally leads to some difficulties.

Example 1: Using the definition of matrix exponential or, if you prefer, the Lagrange-Sylvester formula you can see that

$$\exp \begin{bmatrix} \lambda & 0 \\ 0 & \mu \end{bmatrix} = \begin{bmatrix} e^\lambda & 0 \\ 0 & e^\mu \end{bmatrix}$$

Therefore $\begin{bmatrix} 0 & 0 \\ 0 & 2\pi i \end{bmatrix}$ and $\begin{bmatrix} 0 & 0 \\ 0 & 0 \end{bmatrix}$ are both logarithms of $\begin{bmatrix} 1 & 0 \\ 0 & 1 \end{bmatrix}$.

These difficulties don't arise when we're dealing with real scalars because in that context we only take logarithms of positive numbers τ, and we mean the real logarithm of τ when we write ℓn τ.

How to Find Some of the Logarithms of an Arbitrary Nonsingular Matrix: Special Logarithms

Having chosen a logarithm ℓn(λ), the Taylor series expansion about λ ≠ 0 gives us:

$$\ell n(\tau) = \ell n(\lambda) - \sum_{n=1}^{\infty} \frac{\frac{1}{n}(\lambda - \tau)^n}{\lambda^n} \quad \text{for all } |\tau - \lambda| < |\lambda|$$

With the Lagrange-Sylvester formula in mind, this series leads quite naturally to the following.

Recipe for a Logarithm of the Nonsingular Matrix A:

1. Choose a Jordan form J for A

2. Choose a matrix P which puts A into the Jordan form
    J (i.e., $A = PJP^{-1}$)

3. Choose a logarithm $\ln(\lambda_i)$ for each block $J_i =$
    $\lambda_i I_{r_i} + U_{r_i}$ of J

4. Define

$$\ell(A) = P\left(\overset{m}{\underset{i=1}{\oplus}}\left[\ln(\lambda_i)I_{r_i} - \sum_{n=1}^{r_i-1}\frac{1}{n}\left(\frac{-U_{r_i}}{\lambda_i}\right)^n\right]\right)P^{-1}$$

Now we must see if our recipe works, i.e., we have to
see if $e^{\ell(A)} = A$. To begin with, let's examine the expression
in square brackets in (4):

Let

$$L_i = \ln(\lambda_i)I_{r_i} - \sum_{n=1}^{r_i-1}(\tfrac{1}{n})\left(\frac{-U_{r_i}}{\lambda_i}\right)^n \tag{17}$$

If we can show that $e^{L_i} = J_i$, then we'd be through because

$$e^{\ell(A)} = P\left(\overset{m}{\underset{i=1}{\oplus}} e^{L_i}\right)P^{-1} = P\left(\overset{m}{\underset{i=1}{\oplus}} J_i\right)P^{-1} = PJP^{-1} = A$$

Let $V = U_{r_i}/\lambda_i$ and $W(t) = \sum_{n=1}^{r_i-1}(1/n)(tV)^n$ for all real t.

Straightforward computations show that both $I - tV$ and $e^{-W(t)}$
are solutions of the differential equation $dX/dt = -V(I - tV)^{-1}X$
and both take on the value I at t = 0. Therefore, by theorem
1, $e^{-W(t)} = I - tV$ for all t. In particular, $e^{-W(-1)} = I + V$,

so $\lambda_i e^{-W(-1)} = J_i$. Now

$$\lambda_i e^{-W(-1)} = \left(e^{\ln \lambda_i} I_{r_i} \right) e^{-W(-1)} = e^{(\ln \lambda_i)I_{r_i} - W(-1)} = e^{L_i}$$

so $e^{L_i} = J_i$ for each $1 \le i \le m$. Therefore $e^{\ell(A)} = A$.

Since $\ell(A)$ is a logarithm of A which is produced by the special recipe given above, we'll call it a *special logarithm* for now.

Example 2: Suppose $A = \begin{bmatrix} 0 & 1 & 0 \\ 1 & 0 & 0 \\ 0 & 0 & 1 \end{bmatrix}$. We'll construct two different special logarithms $\ell_1(A)$, $\ell_2(A)$, for $\ell_1(A)$:

1. $J = \begin{bmatrix} 1 & 0 & 0 \\ 0 & -1 & 0 \\ 0 & 0 & 1 \end{bmatrix}$ so $\lambda_1 = 1$, $\lambda_2 = -1$, $\lambda_3 = 1$.

2. $P_1 = \begin{bmatrix} 1 & 1 & 0 \\ 1 & -1 & 0 \\ 0 & 0 & 1 \end{bmatrix}$ (you can verify that $P_1 J P_1^{-1} = A$).

3. Select $\ln \lambda_1 = 0$, $\ln \lambda_2 = i$, $\ln \lambda_3 = 0$.

4. $\ell_1(A) = \dfrac{i\pi}{2} \begin{bmatrix} 1 & -1 & 0 \\ -1 & 1 & 0 \\ 0 & 0 & 0 \end{bmatrix}$.

For $\ell_2(A)$:

Ingredients (1), (2) are the same as $\ell_1(A)$ but we alter the third ingredient by

3. selecting $\ln \lambda_1 = 2\pi i$, $\ln \lambda_2 = i\pi$, and $\ln \lambda_3 = -2\pi i$,

obtaining $\ell_2(A) = \dfrac{i\pi}{2} \begin{bmatrix} 3 & 1 & 0 \\ 1 & 3 & 0 \\ 0 & 0 & -2 \end{bmatrix}$ when we carry out

step (4).

How to Find all the Logarithms of A

We've just seen how to construct certain logarithms of

A -- the special logarithms. How do we find the others?
Fortunately, there aren't any others. This labour-saving fact
is a corollary to the following theorem.

   Theorem 2:   Suppose $P^{-1}AP = J$ is a Jordan form of A.   X
is a logarithm of A iff for some special logarithm $\ell(J)$ of
J and some matrix S commuting with J we have

$$X = PS\ell(J)S^{-1}P^{-1}$$

   Proof: Suppose $J = \overset{m}{\underset{i=1}{\oplus}} J_{r_i}(\lambda_i)$ and X is a logarithm of
A. We want to construct $\ell(J)$ and S with the required properties.

   X is similar to some Jordan matrix K. Suppose K =
$\overset{n}{\underset{j=1}{\oplus}} J_{q_j}(\mu_j)$. We have

$$e^K = \overset{n}{\underset{j=1}{\oplus}} e^{\mu_j} e^{U_{q_j}}$$

$$= \overset{n}{\underset{j=1}{\oplus}} Q_j \left(J_{q_j}(e^{\mu_j}) \right) Q^{-1} \quad \text{(by Exercise 25(c))}$$

$$= Q \left(\overset{n}{\underset{j=1}{\oplus}} J_{q_j}(e^{\mu_j}) \right) Q^{-1} \quad \text{if we set } Q = \overset{n}{\underset{j=1}{\oplus}} Q_j$$

But $e^K \sim e^X$ (because $K \sim X$) and $e^X = A$, therefore $\overset{n}{\underset{j=1}{\oplus}} J_{q_j}(e^{\mu_j})$
is another Jordan form of A, and hence n = m and there is a
permutation $(j_1, j_2, \ldots, j_m)$ of (1, 2, ..., m) such that
$J_{r_i}(\lambda_i) = J_{q_{j_i}}(e^{\mu_{j_i}})$ for i = 1, 2, ..., m. Choose $\ell n\ \lambda_i = \mu_{j_i}$
and, using this determination of the logarithm of λ_i we have

$$\ell(J) = \bigoplus_{i=1}^{m} (\ell n\ \lambda_i) I_{r_i} - \sum_{n=1}^{r_i-1} \left(\tfrac{1}{n}\right)\left(\frac{-U_{r_i}}{\lambda_i}\right)^n$$

Now we'll show that this choice of scalar logarithms produces the required $\ell(J)$:

$$X = R\left(\bigoplus_{i=1}^{m} J_{r_i}(\ell n\ \lambda_i)\right)R^{-1} \quad \text{for some R since X} \sim \text{K}$$

$$= R\left(\bigoplus_{i=1}^{m} V_i \ell[J_{r_i}(\lambda_i)]V_i^{-1}\right)R^{-1} \quad \text{(by Exercise 25(b))}$$

$$= RV\ell(J)V^{-1}R^{-1} \quad \text{if we put } V = \bigoplus_{i=1}^{m} V_i$$

If we define $S = P^{-1}RV$, then $PS\ell(J)S^{-1}P^{-1} = RV\ell(J)V^{-1}R^{-1} = X$ and $SJS^{-1} = P^{-1}RVJV^{-1}R^{-1}P = P^{-1}e^XP = P^{-1}AP = J$.

Conversely, suppose $X = PS\ell(J)S^{-1}P^{-1}$ and $SJ = JS$. We then have $e^X = PSe^{\ell(J)}S^{-1}P^{-1} = PSJS^{-1}P^{-1} = PJP^{-1} = A.$ []

Corollary: Every logarithm of A is a special logarithm of A.

Proof: Suppose $e^X = A$. We want to express X as a special logarithm of A.

1. Let J be a Jordan form of A. Let P be a matrix such that $A = PJP^{-1}$. We now know that for a certain choice of scalar logarithms (one for each block of J) defining a special logarithm $\ell(J)$ of J and a certain matrix S commuting with J, we have $X = PS\ell(J)S^{-1}P^{-1}$.

2. Choose $P_1 = PS$ (you can verify that $A = P_1JP_1^{-1}$).

3. Choose the same scalar logarithms that determined $\ell(J)$.

4. Our recipe then produces the special logarithm

$$\ell_1(A) = PS\ell(J)S^{-1}P^{-1}$$
$$= X$$

and hence X is a special logarithm of A. []

We can now look on our recipe as a way of producing all the logarithms of A [by varying (1), (2) and (3)] so we needn't refer to these $\ell(A)$ as "special" logarithms any longer and we'll drop the notation $\ell(A)$ in favour of the traditional $\ell n\ A$.

Exercises

25. Suppose $U = J_r(0)$, $\lambda \neq 0$, $W = \sum_{n=1}^{r-1} (-U/\lambda)^n(1/n)$, and

$L = (\ell n\ \lambda)I - W$ where $\ell n\ \lambda$ is an arbitrary logarithm of λ.

a. Show that W is similar to U. (Suggestion: find the index of nilpotence (see Chap. 1, Exercise 6) of W and consider what it implies for the Jordan form of W.

b. Show that L is similar to $J_r(\ell n\ \lambda)$ and hence that

c. $J_r(\lambda)$ is similar to λe^U.

26. What are the possible values of $\ell n\ I$?

27. Prove that every logarithm of A commutes with A.

28. a. If A commutes with B and $\ell n\ A$, $\ell n\ B$ are arbitrary logarithms of A and B respectively, prove that $\ell n\ A + \ell n\ B$ is a logarithm of AB.

b. Find 2×2 examples of $\ell n\ A$, $\ell n\ B$, $\ell n\ AB$, such that $AB = BA$ but $\ell n\ A + \ell n\ B \neq \ell n\ AB$.

c. Find 2 × 2 examples of $\ln A$, $\ln B$ such that $\ln A +$
$\ln B \neq \ln AB$ for any $\ln AB$. (Suggestion: Try
$A = \begin{bmatrix} 0 & 1 \\ 1 & 0 \end{bmatrix}$, $B = \begin{bmatrix} 1 & 1 \\ 0 & 1 \end{bmatrix}$, choose $\ln A$ so that $\ln A +$
$\ln B$ is singular; show that every logarithm of AB
is nonsingular.)

29. a. If n is any integer and $\ln A$ is any logarithm of A
show that $n(\ln A)$ is a logarithm of A^n. (Suggestion:
Use 28(a) and begin with $n > 0$.)

b. Find a (nonsingular, of course) matrix A and logarithms
$\ln A$ and $\ln A^2$ such that $2(\ln A) \neq \ln A^2$.

30. a. Suppose $J = \overset{m}{\underset{i=1}{\oplus}} J_{r_i}(\lambda_i)$ is a Jordan matrix and
$\lambda_i \neq \lambda_j$ when $i \neq j$. If $SJ = JS$ prove that $S = \overset{m}{\underset{i=1}{\oplus}} S_i$
where each S_i commutes with $J_{r_i}(\lambda_i)$ (see Chap. 1,
Exercise 31).

b. Suppose the minimal polynomial of A is the characteristic
polynomial of A and $P^{-1}AP = J$ is a Jordan form of A.
Prove that X is a logarithm of A iff $X = P\ell(J)P^{-1}$
for some special logarithm of $\ell(J)$ of J.

Arbitrary Powers of a Nonsingular Matrix

With the aid of the matrix logarithm we can define
arbitrary powers (such as A^π or $A^{(1+i\sqrt{2})}$) for any nonsingular
A by fixing one logarithm of A call it $\ln A$ and letting (1)
$A^\sigma = e^{\sigma \ln(A)}$ for all scalars σ. Notice that A^σ usually
depends on the choice of $\ln A$ as well as on σ, but $A^0 = I$
for any logarithm defining A^0, since $e^{0 \ln A} = e^0 = I$ for all
$\ln A$.

Exercises

31. Show that no matter how $\ln A$ is chosen to define A^σ, if σ is a positive integer, then $A = \underbrace{AA\cdots A}_{\sigma \text{ factors}}$

    and hence (1) is consistent with our earliest definition of integral matrix powers.

32. Suppose we fix one logarithm to define Λ^σ for all σ.

    a. Show that $A^\sigma A^\tau = A^{\sigma+\tau}$.

    b. Show that $A^{-\sigma}$ is the inverse of A^σ.

    c. Show that $A^{\sigma n} = (A^\sigma)^n$ for all integers n.

    d. To what extent is it true that $A^{\sigma n} = (A^n)^\sigma$ when n is an integer? [Think about how A^σ is defined and how B^σ is defined $(B = A^n)$.]

    e. To what extent is it true that $A^{\sigma\tau} = (A^\sigma)^\tau$?

33. Suppose a particular $\ln A$ is used to define A^σ (all σ).

    a. Prove that $d/d\sigma(A^\sigma) = (\ln A)A^\sigma$, and hence that

    b. $d/dt(A^{\sigma t}) = [\sigma\, \ln(A)]A^{\sigma t}$ for all σ, t

Rational Powers of Singular Matrices

If $X^2 = A$, we call X a *square root of* A. We've seen that nonsingular matrices have square roots (if A is nonsingular, then, choosing a value of $A^{1/2}$, we have $(A^{1/2})^2 = A$ by Exercise 32(c). If A is singular, then A may or may not have square roots. For instance $\begin{bmatrix} 1 & 0 \\ 0 & 0 \end{bmatrix} = \begin{bmatrix} 1 & 0 \\ 0 & 0 \end{bmatrix}^2 = \begin{bmatrix} -1 & 0 \\ 0 & 0 \end{bmatrix}^2$ but $\begin{bmatrix} 0 & 1 \\ 0 & 0 \end{bmatrix} \neq X^2$ for any X (see Exercise 34(b)).

Exercises

34. Suppose A is nilpotent. Show that

   a. Every square root of A is nilpotent;

   b. If $m \geq 1$, then $J_{2m}^2(0) \sim J_m(0) \oplus J_m(0)$ and
       $J_{2m+1}(0) \sim J_{m+1}(0) \oplus J_m(0)$;

   c. If $X^2 = A$, A has a_j Jordan blocks of order j and X
       has x_j Jordan blocks of order j (all $j \geq 1$), then

   $$a_j = x_{2j+1} + 2x_{2j} + x_{2j-1} \quad \text{for } j = 1, 2, \ldots, k \tag{1}$$

   (Note: $a_j = x_j = 0$ for $j \geq k$);

   d. If $a_1, a_2, \ldots, a_j, \ldots$ are as in (c) and there
       exist nonnegative integers x_j satisfying Eq. (1)
       of (c), then A has a square root. [Suggestion: If
       $P^{-1}AP = J$ is a Jordan form of A and $\hat{J}$ is a Jordan
       matrix having x_j blocks of order j for $1 \leq j \leq k$,
       show that $J^2 \sim J$. It then follows that $A = Q\hat{J}^2Q^{-1}$
       for some Q (why?), and hence $A = (Q\hat{J}Q^{-1})^2$.]

35. Suppose A is singular. It follows that $A = P(N \oplus M)P^{-1}$
     for some P, where N is nilpotent and M is nonsingular
     ($A = PNP^{-1}$ if A is nilpotent). If N has a square root,
     denote it by S. Since M is nonsingular, it has a square
     root, choose one and call it T. We then have

   $$[P(S \oplus T)P^{-1}]^2 = A$$

   a. Show that a singular matrix A has a square root iff
       there exist nonnegative integers x_j satisfying (1)
       of 34(c) when a_j denotes the number of Jordan blocks
       of A of order j for the eigenvalue 0.

(Notice that the square root $P(S \oplus T)P^{-1}$ of A which we
constructed depended on the choices of Jordan form, P
and T. Call a square root constructed that way a
special square root. Are there any square roots other
than special square roots?)

    b. Carefully reexamine the proof of theorem 2 and its
       corollary to see if the methods used there can be
       modified to determine whether all square roots are
       special.

36. We say that X is a *cube root of A* iff X^3 = A. If A is
    nilpotent, find conditions analogous to (1) of 34(c)
    which are necessary and sufficient for A to have a
    cube root. Prove an analogue of Exercise 35(a) for
    cube roots of singular matrices A and see, as in Exercise
    35(b) if your construction of a cube root can produce
    all the cube roots of A.

37. If A is singular and r is a positive rational number, how
    would you define A^r? To what extent would it then be
    true that $(A^r)^s = A^{rs}$ when r and s are positive rationals?

8. STABILITY

Many physical systems are governed by a differential equation
of the form:

$$\frac{dx}{dt} = A(t)x(t)$$
$$x(0) = c \qquad (t \geq 0) \qquad (18)$$

The condition of the system at time $t \geq 0$ is described by
the function x(t) = X(t)c, where X is the fundamental matrix

of $dX/dt = A(t)X(t)$. The function X depends on c as well as
t. The behaviour of the physical system can be studied by
investigating the influence of the initial conditions c on
the solution function x. For that reason we shall now change
our notation slightly and define $x_c(t) = X(t)c$ to emphasize
the connection between the solution and the initial conditions.
In particular, x_0 (= the zero function) is the (only) solution
to Eq. (18) with $x(0) = 0$; x_0 is also referred to as *the zero
solution*.

It may be useful to imagine that c represents the k
values of k inputs into a physical system governed by Eq. (18)
and that $x_c(t)$ describes the condition of k outputs at time
$t \geq 0$. Think of the inputs $c_1,\ldots, c_k$ as settings of k
control knobs and the outputs as readings of k meters, the
i^{th} meter registers the ith entry in $x_c(t)$ at time t.

The zero solution to Eq. (18) is said to be *semi-stable*
iff $\lim_{c \to 0} x_c(t) = 0$ uniformly in t, i.e., for any $\epsilon > 0$ there
is a δ_ϵ such that for all $t \geq 0$,

$$||x_c(t)|| \leq \epsilon \quad \text{if} \quad ||c|| \leq \delta_\epsilon$$

Note that δ depends at most on ϵ and not at all on t. Thus
when x_0 is semi-stable, a small change in the settings on
the control knobs away from 0 will result in only small
deflections of the meter needles. These deflections may
persist for all time but they will always be small.

The zero solution to Eq. (18) is said to be *stable* iff
$\lim_{t \to +\infty} x_c(t) = 0$ for all c. Thus when x_0 is stable, no matter

how the dials are set, all the meters will read zero
eventually.

The zero solution to Eq. (18) is said to be *unstable* iff
it is not semi-stable.

Please note that in more general settings than ours
[where the equation $dx/dt = f(x)$ is studied instead of the
special case where $f(x) = A(t)x(t)$] the definition of "stable"
is: x_0 is semi-stable and, for some $\delta > 0$, if $||c|| < \delta$, then
$\lim_{t \to +\infty} x_c(t) = 0$. In our special case this seemingly more
generous requirement that $\lim_{t \to +\infty} x_c(t) = 0$ only need hold for
all c such that $||c|| < \delta$ is really equivalent to the condition
that "$\lim_{t \to +\infty} x_c(t) = 0$ for all c" and, as we shall show shortly,
in our setting, the condition "$\lim_{t \to +\infty} x_c(t) = 0$ for all c"
implies that x_0 is semi-stable.

Two physical examples are given in Sec. 12. You might
find it useful to examine them now.

Example 1: The zero solution to $dx/dt = x$ is unstable
even though for every $t \geq 0$ and every $\varepsilon > 0$ there is a δ
(namely $e^{-t}\varepsilon$) such that $||x_c(t)|| \leq \varepsilon$ if $||c|| \leq \delta$.

Example 2: The zero solution to $dx/dt = (\sin t)x$ is
semi-stable as $||e^{-\cos t}c|| \leq e||c||$ for all t, but it's not
stable because $e^{-\cos t}$ has no limit as $t \to +\infty$.

Example 3: The zero solution to $dx/dt = -x$ is stable
because $||x_c(t)|| = ||e^{-t}c|| = e^{-t}||c|| \to 0$ as $t \to +\infty$ for all
c.

Exercises

38. Show that the zero solution in Example 1 is unstable.

39. If M(t) is any matrix valued function and $\lim_{t \to +\infty} M(t)c = 0$

    for all constant vectors c, show that

$$\lim_{t \to +\infty} M(t) = 0$$

Lemma: Stability implies semi-stability.

Proof: Suppose the zero solution to dx/dt = Ax is stable and X is the fundamental matrix for A. Since $\lim_{t \to +\infty} ||x_c(t)|| = 0$ for all c (because x_0 is stable) and $x_c(t) = X(t)c$, Exercise 39 implies that $\lim_{t \to +\infty} X(t) = 0$, and hence that $||X(t)|| \leq 1$ for all t sufficiently large, say for all $t \geq \tau$. If we let $\beta = \text{lub}\{||X(t)|| : 0 \leq t \leq \tau\}$, then $||X(t)|| \leq \beta + 1$ for all $t \geq 0$. Let $\varepsilon > 0$. If $||c|| \leq \varepsilon/(\beta + 1)$, then $\varepsilon \geq (\beta + 1)||c|| \geq ||X(t)||\,||c|| \geq ||X(t)c||$ for all $t \geq 0$. Therefore x_0 is semi-stable. []

Exercise

40. Suppose X is the fundamental matrix for A(t). Show that the zero solution to dx/dt = A(t)x is

    a. semi-stable iff "X(t) is bounded on $t \geq 0$", i.e., there is a number γ such that for all $t \geq 0$,

    $||X(t)|| \leq \gamma$

    b. stable iff $\lim_{t \to \infty} X(t) = 0$

    If you didn't use the Lemma to prove (a) or (b), then (a) and (b) provide an alternative proof that stability implies semi-stability, because:

c. If M(t) is continuous at each $t \geq 0$ and $\lim\limits_{t \to \infty}$ M(t) = 0,

is bounded on $t \geq 0$.

9. STABILITY WHEN A IS CONSTANT

When A is constant, the solution to dx/dt = Ax(t) subject to
x(0) = c becomes $x_c(t) = e^{At}c$ for all $t \geq 0$ (as we have seen
before).

Lyapunov's Criterion: If A is constant, then the zero
solution to dx/dt = Ax is stable iff the real part of each
eigenvalue of A is negative.

Proof: Suppose $Re(\lambda) < 0$ for all eigenvalues λ of A.
Exercise 20 implies that $\lim\limits_{t \to \infty} e^{At} = 0$, and hence $\lim\limits_{t \to \infty} x_c(t) = 0$
for all constant vectors c.

We'll prove the converse. Assume x_0 is stable. Therefore

$$\lim_{t \to \infty} e^{At}c = 0 \quad \text{for all c} \tag{19}$$

As in Sec. 6 we have

$$P\,e^{At}P^{-1} = \overset{m}{\underset{\ell=1}{\oplus}} e^{\lambda_\ell t}\left(\sum_{j=0}^{n_\ell - 1} \frac{(tU_{n_\ell})^j}{j!} \right)$$

Now $e^{\lambda_\ell t} = e^{\alpha_\ell t}(\cos \beta_\ell t + i \sin \beta_\ell t)$, where $\alpha_\ell = Re(\lambda_\ell)$ and
$\beta_\ell = Im(\lambda_\ell)$. Inasmuch as

$$||Pe^{At}P^{-1}|| = \sum_{\ell=1}^{m} \sum_{j=0}^{n_\ell - 1} \left| e^{\lambda_\ell t} \right| \frac{t^j}{j!} \geq \left| e^{\lambda_\ell} \right| (t^j/j!)\frac{t^j}{j!}$$

and

$$\left| e^{\lambda_\ell t} \right| = e^{\alpha_\ell t}$$

we have

$$(k!) \left| \left| Pe^{At}P^{-1} \right| \right| \geq e^{\alpha_\ell t} \geq 0 \qquad (20)$$

for all ℓ and all $t \geq 1$. Equation (19) implies that $\lim_{t \to \infty} e^{At} = 0$ (using Exercise 38), and hence that $\lim_{t \to \infty} Pe^{At}P^{-1} = 0$. Therefore (20) implies that $\lim_{t \to \infty} e^{\alpha_\ell t} = 0$, and hence $\alpha_\ell < 0$. □

Exercise

41. a. State a condition on the eigenvalues of A which is equivalent to the semi-stability of x_0 when A is constant.

b. Prove the statement made in (a).

A constant matrix A is called a *stability matrix* iff the real part of each of its eigenvalues is negative. Lyapunov's Criterion can then be put this way: When A is constant, the zero solution to $dx/dt = Ax$ is stable iff A is a stability matrix.

In practice only small matrices can be tested for stability by directly finding the eigenvalues of A. Larger matrices require indirect methods.

10.   INDIRECT TESTS FOR STABILITY: LYAPUNOV'S THEOREM

If M is a complex k × ℓ matrix, M^* is the ℓ × k matrix whose ij^{th} entry is the complex conjugate of m_{ji}. M^* is known as the *conjugate-transpose* of M. You can easily verify that $(MN)^* = N^*M^*$, whenever the product MN is defined, and that $M^{**} = M$.

If $M^* = M$, then M is said to be *Hermitian*. A matrix M is said to be *symmetric* iff $M = M^{tr}$. Thus real Hermitian matrices are symmetric and real symmetric matrices are Hermitian.

If H is a Hermitian matrix and x is any complex vector, then x^*Hx is real because the conjugate of the complex number x^*Hx is $(x^*Hx)^* = x^*H^*x^{**} = x^*Hx$, the number itself. If $x^*Hx > 0$ for all nonzero complex vectors x, then H is said to be *positive definite*. The identity matrix is an example of a positive definite matrix because

$$x^*Ix = x^*x = \sum_{i=1}^{k} x_i^* x_i$$

$$= \sum_{j=1}^{k} |x_j|^2 \quad \text{(as } x_j^* \text{ is the complex conjugate of } x_j\text{)}$$

$$> 0 \quad \text{iff} \quad x \neq 0$$

In 1892 Lyapunov proved that a real matrix A is a stability matrix iff there exists a unique real positive definite symmetric matrix V such that

$$A^{tr}V + VA = -I \tag{21}$$

Before we prove this theorem we ought to point out why the test for stability inherent in it can be superior to a direct examination of the real parts of the eigenvalues of A (when

k is large). Such a direct examination, of course, requires

finding all the eigenvalues of A. When k is large that is

usually harder to do than solving Eq. (21) for V and (in the

event that there is a unique solution) seeing if it is

symmetric and positive definite. Solving Eq. (21) consists

of solving a linear inhomogeneous system of k^2 equations in

the k^2 unknowns v_{11}, v_{12}, ..., v_{ij}, ..., v_{kk}. Tests for

positive definiteness will be discussed later on.

Exercise

42. If M is a real symmetric matrix and $x^{tr}Mx > 0$ for all

     real nonzero vectors x, show that $x^*Mx > 0$ for all

     complex nonzero vectors x.

Lemma: Suppose A and V are constant matrices, c is a

constant complex vector and $x(t) = e^{At}c$ for all real t. If

$f(t) = x^*Vx$, then $df/dt = x^*(A^*V + VA)x$.

   Proof:

$$\frac{df}{dt} = \frac{d(x^*)}{dt} Vx + x^*V \frac{dx}{dt} \quad (\text{Exercise } 2(b))$$

$$= x^*A^*Vx + x^*VAx \quad \text{as} \quad \frac{dx^*}{dt} = \left(\frac{dx}{dt}\right)^* = (Ax)^* = x^*A^*$$

$$= x^*(A^*V + VA)x \quad \square$$

Lyapunov's Theorem (1892): The constant real matrix A

is a stability matrix iff there exists a unique real positive

definite symmetric matrix V satisfying

$$A^{tr}V + VA = -I \qquad\qquad (21)$$

Proof: Suppose A is a real stability matrix. Let $F(M) = A^{tr}M + MA$ for every square real matrix M. F is a linear transformation of the vector space $\underline{R}^{k^2}$ of all square real matrices into itself. Let's examine[†] the nullspace of F: Suppose $F(Z) = 0$, that is, $-A^{tr}Z = ZA$. It follows from Exercise 43 that $e^{-tA^{tr}}Z = Ze^{At}$. In Chap. 3, Sec. 6 we saw that $e^{A+B} = e^A e^B$ when $AB = BA$, therefore $e^B e^{-B} = e^{B-B} = e^0 = I$ so $e^{-B} = (e^B)^{-1}$. Thus

$$Z = e^{tA^{tr}} Ze^{At} \quad \text{for all } t \qquad\qquad (22)$$

But both A and A^{tr} are stability matrices (A^{tr} is a stability matrix because its eigenvalues are the same as those of the stability matrix A) so $\lim_{t\to\infty} e^{tA^{tr}} = \lim_{t\to\infty} e^{At} = 0$; therefore (taking the limit as $t \to \infty$ in Eq. (22)) we have $Z = 0$. This shows that the nullspace of F consists of just the matrix 0. As $\underline{R}^{k^2}$ is finite dimensional, this implies that F is invertible. So, if we put $V = F^{-1}(-I)$, we have $F(V) = -I$, that is,

$$A^{tr}V + VA = -I \qquad\qquad (21)$$

No other matrix V satisfies Eq. (21), for if $A^{tr}W + WA = -I$, then $F(W) = -I$; so $W = F^{-1}(-I) = V$.

[†] We could now invoke Chap. 1, lemma 7.1 to show that F is nonsingular (see Chap. 1, Exercises 31, 32), but we'll present an alternate proof in case you've not gone over the material.

$$F(V^{tr}) = A^{tr}V^{tr} + V^{tr}A = (VA)^{tr} + (A^{tr}V)^{tr}$$

$$= (VA + A^{tr}V)^{tr} = (-I)^{tr} = -I$$

Therefore V^{tr} satisfies Eq. (21); so $V^{tr} = V$, and hence V is symmetric. As $V \in \underline{R}^{k^2}$ we have shown so far that Eq. (21) has a unique real symmetric solution. It only remains to show that V is positive definite, i.e., that $c^{tr}Vc > 0$ for all nonzero c; we do so by contradiction: Suppose $c^{tr}Vc \leq 0$ and $0 \neq c \in \underline{R}^k$. Let $x = e^{At}c$ and $f(t) = x^{tr}Vx$. The previous lemma and Eq. (21) imply that $df/dt = -x^{tr}x$, and hence (as $x^{tr}x > 0$) $df/dt < 0$ for all $t \geq 0$. This means that f is a monotone decreasing function on $t \geq 0$. Consequently,

$$f(1) < f(0) = c^{tr}Vc \leq 0$$

Therefore

$$f(t) \leq f(1) < 0 \text{ for all } t \geq 1 \qquad\qquad (23)$$

On the other hand $\lim\limits_{t \to \infty} f(t) = 0$ because A is a stability matrix which contradicts Eq. (23). Therefore V is positive definite.

To prove the converse, assume Equation (21) is satisfied by a real symmetric positive definite matrix V. Let λ be any eigenvalue of A. We then have $Av = \lambda v$ for some nonzero complex v. Let $f(t) = (e^{\lambda t}v)^{*}Ve^{\lambda t}v$. Differentiating we find that $f'(t) = \lambda^{*}f(t) + \lambda f(t)$ (writing f' for df/dt), and hence

$$f'(t) = 2 \, \text{Re}(\lambda)f(t) \quad \text{for all } t \tag{24}$$

On the other hand, $(At)^n/(n!) \, v = (\lambda t)^n/(n!) \, v$ for all n,
so $e^{At}v = e^{\lambda t}v$. Therefore $f(t) = (e^{At}v)^* V e^{At}v$. Applying
the lemma we obtain

$$f'(t) = -(e^{At}v)^* e^{At}v = -v^* e^{\lambda^* t} e^{\lambda t}v = -e^{2\text{Re}(\lambda)t} v^* v$$

and hence $f'(0) = -v^* v$, therefore

$$-v^* v = f'(0) = 2 \, \text{Re}(\lambda)f(0) = 2 \, \text{Re}(\lambda)(v^* V v)$$

But $v^* V v > 0$ and $v^* v > 0$ because V and I are positive definite.
Therefore $\text{Re}(\lambda) < 0$. Since λ was an arbitrary eigenvalue
of A, we have shown that A is a stability matrix. []

Exercises

43. If $f(\tau) = \sum\limits_{n=0}^{\infty} \alpha_n \tau^n$ has a radius of convergence ρ,

     $|C| < \rho$, $|B| < \rho$ and $CV = VB$, show that $f(C)V = Vf(B)$.

44. If A and B are $k \times k$ real stability matrices and C is
     any $k \times k$ real matrix, show that the equation $AM + MB = C$
     has one, and only one, solution M. Use the method
     given in the proof of Lyapunov's theorem.

This theorem reduces the problem of testing A for
stability to (1) solving Eq. (21) for V and (if there is a
unique solution and it's symmetric) (2) testing V to see
if it's positive definite.

Solving Eq. (21) involves the routine solution of a system of k^2 linear equations in k^2 unknowns.

Various methods are available for (2) [see, e.g., Gantmacher (1959, vol. I, pp. 299-308)]. For instance, if

$$v_{11} > 0, \det\begin{bmatrix} v_{11} & v_{12} \\ v_{21} & v_{22} \end{bmatrix} > 0, \ldots, \det\begin{bmatrix} v_{11} & \cdots & v_{1j} \\ \vdots & & \vdots \\ v_{j1} & \cdots & v_{jj} \end{bmatrix} > 0, \ldots,$$

det V > 0 then V is positive definite. It turns out [see Gantmacher (1959, vol. II, pp. 225-231)] that this sequence of inequalities for V is related to another test for the stability of A.

Routh-Hurwitz Criterion: Suppose A is a real matrix whose characteristic polynomial is $\tau^k + \alpha_1 \tau^{k-1} + \alpha_2 \tau^{k-2} + \cdots + \alpha_k$, then A is a stability matrix iff

$$\alpha_1 > 0, \det\begin{bmatrix} \alpha_1 & \alpha_3 \\ 1 & \alpha_2 \end{bmatrix} > 0, \quad \det\begin{bmatrix} \alpha_1 & \alpha_3 & \alpha_5 \\ 1 & \alpha_2 & \alpha_4 \\ 0 & \alpha_1 & \alpha_3 \end{bmatrix} > 0, \ldots,$$

$$\det\begin{bmatrix} \alpha_1 & \alpha_3 & \alpha_5 & \alpha_7 & \cdots \\ 1 & \alpha_2 & \alpha_4 & \alpha_6 & \cdots \\ 0 & \alpha_1 & \alpha_3 & \alpha_5 & \cdots \\ 0 & 1 & \alpha_2 & \alpha_4 & \alpha_5 \\ 0 & 0 & \alpha_1 & \alpha_3 & \alpha_5 \\ \vdots & & & & \ddots \\ 0 & 0 & \cdots & & \alpha_k \end{bmatrix} > 0$$

where $\alpha_\ell = 0$ if $\ell > k$.

Of course, the usefulness of this criterion is limited by the difficulties of calculating determinants of large matrices.

You have seen that even though the question: "When is
the zero solution to dx/dt = Ax(t) stable?" has a straight
forward enough answer if A is constant ("When the real parts
of its eigenvalues are negative"), the usefulness of the
answer may be limited by our ability to tell when the
eigenvalues of a given matrix have this property. In Chap.
4 we will have a brief look at the general question of finding
(small) regions of the complex plane containing the eigenvalues
of a given matrix A.

11. LYAPUNOV TRANSFORMATIONS: STABILITY WHEN A(t) IS PERIODIC

Let's return to the problem of determining the stability of
the zero solution to dx/dt = A(t)x(t). If we put x = L(t)y
and d = $L^{-1}(0)c$, then

$$dx/dt = A(t)x(t) \quad x(0) = c \qquad\qquad (25)$$

iff

$$dy/dt = B(t)y(t) \quad y(0) = d \qquad\qquad (26)$$

where B = $L^{-1}AL - L^{-1} dL/dt$ as you can verify by direct
computation.

We would like to find a change of variable matrix L(t)
which would make (26) an equation whose zero solution has
the same stability character as that of (25) but whose
stability is easier to test than that of (25). For instance,
if we could find L so that B is constant, we would use the
results of Sec. 10 to determine the stability of the zero
solution to dx/dt = A(t)x(t) even though A isn't constant.

L is called a *Lyapunov matrix* iff for all t $\geq$ 0,

1. dL/dt is continuous

2. L and dL/dt are bounded

3. glb{|det L(t)| : t $\geq$ 0} > 0

Exercise

45. Show that L^{-1} is a Lyapunov matrix if L is a Lyapunov matrix.

Theorem 3: If L is a Lyapunov matrix, then the zero solution to (25) is

1. Stable iff the zero solution to Eq. (26) is stable

2. Semi-stable iff the zero solution to Eq. (26) is semi-stable

3. Unstable iff the zero solution to Eq. (25) is unstable

Proof: We have x = L(t)y and c = L(0)d. Suppose y_d is the solution to Eq. (26) such that y(0) = d, then x is the solution to (25) such that x(0) = c, that is x = x_c. According to (2) in the definition of Lyapunov matrix there is a β > 0 such that ||L(t)|| $\leq$ β for all t $\geq$ 0. We then have

$$||x_c(t)|| = ||L(t)y_d(t)|| \leq ||L(t)||\,||y_d(t)|| \leq \beta||y_d(t)||$$

$$(27)$$

Suppose the zero solution to Eq. (26) is stable, then $\lim_{t \to +\infty} y_d(t)$ = 0, and hence [using (27)] $\lim_{t \to +\infty} x_c(t)$ = 0. Therefore the zero solution to Eq. (25) is stable. L^{-1} is a Lyapunov matrix too (Exercise 45), y = $L^{-1}(t)x$ and d = $L^{-1}(0)c$;

so by the same argument, the zero solution to Eq. (26) is
stable if the zero solution to Eq. (25) is stable. This
establishes (1).

Suppose the zero solution to (26) is semi-stable. Let
$\epsilon > 0$, then there is a δ_ϵ such that for every $t \geq 0$ if
$||d|| < \delta_\epsilon$, then $||y_d(t)|| \leq \epsilon$. Let $\delta'_\epsilon = \delta_{\epsilon/\beta}||L^{-1}(0)||^{-1}$.
If $||c|| \leq \delta'_\epsilon$, then $||d|| = ||cL^{-1}(0)|| \leq ||c||||L^{-1}(0)|| \leq$
$\delta_{\epsilon/\beta}$. Therefore, for all $t \geq 0$ we have $||y_d(t)|| \leq \epsilon/\beta$,
and hence [applying (27)] $||x_c(t)|| \leq \epsilon$. Consequently,
$\lim_{c \to 0} x_c(t) = 0$ uniformly on $t \geq 0$, i.e., the zero solution to
(25) is semistable. Again as in (1) we obtain the converse
by applying Exercise 45. This establishes (2), and hence
(3). []

A matrix valued function P defined on $t \geq 0$ is said to
be *periodic* iff there is a $\tau > 0$ such that $P(t + \tau) = P(t)$
for all $t \geq 0$. The number τ is called a *period* of P. For
example, $P(t) = e^{Bt}$ is periodic if $Re(\lambda) = 0$ for all eigenvalues λ of a diagonable matrix B, in which case 2π is a
period for $P(t)$.

We want to examine the stability of the zero solution
to (25) $dx/dt = P(t)x(t)$ assuming that P is periodic and
continuous. We'll show how to construct a Lyapunov matrix
$L(t)$ which will transform (25) into (26) $dy/dt = Ay$, where
A is a certain *constant* matrix. The previous theorem will
then enable us to give necessary and sufficient conditions
for the stability of the zero solution to $dx/dt = P(t)x(t)$ in
terms of the eigenvalues of the constant matrix A.

Suppose $X(t)$ is the solution to $dX/dt = P(t)X(t)$ such that $X(0) = I$. We know that there is a $\tau > 0$ such that $P(t + \tau) = P(t)$ for all $t \geq 0$.

Let us put $X(\tau) = C$ to remind us that $X(\tau)$ is constant. Jacobi's identity implies that C is nonsingular, so we may define $\ln C$ by means of the Lagrange-Sylvester interpolation formula (see Sec. 7). We can then define C^s for all scalars s by $C^s = e^{s \ln C}$ (before Sec. 7 we only defined C^s for integral s). We define L by:

$$L(t) = X(t)C^{-t/\tau} \quad \text{for all } t \geq 0$$

In Exercise 48 you will see why L is a Lyapunov matrix. Now when we make the substitution $y(t) = L^{-1}(t)x(t)$, we transform the differential equation (25) $dx/dt = P(t)x$ into (26) $dy/dt = [(1/\tau)\ln C]y$ because if x satisfies (25), then $x = Xc$; so $y = L^{-1}Xc = C^{t/\tau}c$, and hence

$$dy/dt = ((1/\tau)\ln C)C^{t/\tau}c \quad \text{[by Exercise 33(b)]}$$

$$= ((1/\tau)\ln C)y$$

Now the zero solution to (25) is stable

iff the zero solution to (26) is stable

iff $(1/\tau)\ln C$ is a stability matrix

iff $\text{Re}[(\ln \gamma)/\tau] < 0$ for every eigenvalue γ of C

iff $\text{Re}(\ln \gamma) < 0$ for every eigenvalue γ of C

iff $\ln|\gamma| < 0$ for every eigenvalue γ of C

iff $|\gamma| < 1$ for every eigenvalue γ of C

Consequently the zero solution to dx/dt = Px is stable
when P is periodic with period τ iff the spectral radius of
X(τ) is less than 1 (where X is the fundamental matrix of
P(t)).

Exercises

46.   Show that if Q(t) is periodic and continuous on t ≥ 0,
      then Q(t) is bounded on t ≥ 0.

47.   Explain why B^s is continuous for all s.

48.   Suppose X, L, C and τ are as in the preceding discussion.

      a.   Show that dL/dt is continuous.

      b.   Show that X(t + τ) = X(t)X(τ) for all t ≥ 0 [show
           both sides satisfy dY/dt = PY subject to Y(0) = X(τ)],
           and hence that both L and dL/dt are periodic.
           (Therefore by Exercise 46, they are both bounded on
           t ≥ 0.)

      c.   Show that glb{|det L(t)| : t ≥ 0} > 0.   (*Hint:* Use
           the fact that L is periodic, continuous, and non –
           singular on t ≥ 0.)

49.   Show that the product of Lyapunov matrices is a Lyapunov
      matrix.   (This exercise together with Exercise 45 shows
      that the Lyapunov matrices are a group under matrix
      multiplication.)

50.   Is $e^{(sin\ t)A}$ a Lyapunov matrix if A ≠ 0 is constant?
      Explain.

12. TWO PHYSICAL EXAMPLES OF STABILITY[†]

    Example 1: The Motion of a Satellite about a Triangular
               Libration Point[††]

In 1772 Lagrange published a memoir entitled *Essai sur*
le Problème des Trois Corps, in which he studied a three-body
problem in which two massive bodies m_1 and m_2 move in circular
orbits about their centre of mass, and a third body m_0
(possibly a satellite) of negligible mass moves about them
under their combined attraction. Negligible mass means one
so small that it does not disturb the two larger masses. In
this memoir Lagrange predicted the existence of five equilibrium
points in the vicinity of the two large bodies. These points
are now called *libration* points. When the small body is
placed at any one of these equilibrium points with zero
relative velocity, it remains there indefinitely. Three of
the libration points (L_1, L_2, L_3) lie on the line passing
through the two massive bodies, the actual location depending
on the mass ratio of the two large bodies. Libration points
L_4 and L_5 occur at the vertices of two equilateral triangles
both lying in the plane of motion of the two massive bodies
and having as a common base the line joining the two massive
bodies.

[†] This material was kindly provided by Professor R.M. Erdahl.

[††] For a more detailed discussion of this material, see
Whittaker (1949, pp. 406-420).

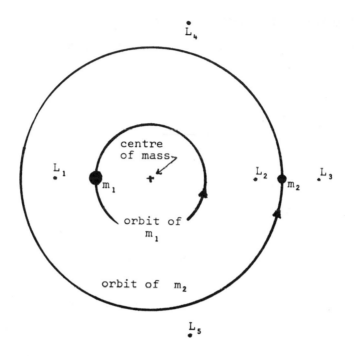

If a satellite or small body is inserted into the
neighbourhood of a libration point with nonzero relative
velocity, as opposed to insertion precisely into the point
with zero relative velocity, will the resulting trajectory
remain close to the libration point for all time, or will the
object depart from the point, never to return? This is a
question regarding the stability properties of Lagrange's
five special solutions of the equations of motion.

Any one of these solutions (and the libration point) is
said to be *stable* if when a satellite is given an initial
position close to that of the libration point and an initial
relative velocity close to zero, it moves progressively
towards the libration point with increasing time. It is said

to be *unstable* if under the same circumstances, the satellite

wanders away from the libration point, and is said to be

semi-stable if it remains in the vicinity but does not move

towards the libration point with increasing time. It turns

out that the three libration points L_1, L_2, L_3 are unstable

and that the two triangular points L_4 and L_5 are unstable if

the mass ratio of the smaller to the larger massive body

exceeds 0.0385. When the mass ratio is less than this figure,

there is some evidence, which we will give below, that the

triangular points are semi-stable.[†]

It is interesting that Lagrange's memoir was published

at a time when only six of the planets were known and none

of the minor planets had been discovered. Thus there was no

evidence that the triangular configuration existed in our

solar system. However in 1906 more than 100 years after

Lagrange's paper, the Konigstul Observatory in Heidelburg

reported the discovery of a minor planet located at

approximately the L_4 position of the Sun-Jupiter system. In

rapid succession 14 additional bodies were discovered in the

two triangular points of the Sun-Jupiter system. The

asteriods in the L_4 position were named after the Greek heroes

of the Trojan war, those at L_5 after the Trojan heroes. The

ratio of the lunar mass to that of the earth is 0.012 and

therefore less than the critical value 0.0385. Though there

[†] See Schutz (1966) for (numerical) solutions of the equations
of motion of a satellite near a triangular libration point.
He also provides additional historical material and other
references regarding this problem.

are no large bodies in the triangular points of the earth-moon
system, there are collections of dust.

If we assume that the two massive bodies are moving in
circular orbits about their centre of mass and if we adopt
a coordinate system:

1.   Whose origin is at the centre of mass of the two
     massive bodies

2.   Rotates with such a constant angular frequency, ω,
     that relative to this frame the two massive bodies
     appear fixed

3.   Is such that the two massive bodies lie on the
     x-axis.

then the equations of motion of the third body are given by

$$\frac{d^2x}{dt^2} - 2\omega \frac{dy}{dt} = \omega^2 x - \mu_1 \frac{(x - x_1)}{r_1^3} - \mu_2 \frac{(x - x_2)}{r_2^3}$$

$$\frac{d^2y}{dt^2} + 2\omega \frac{dx}{dt} = \omega^2 y - \mu_1 \frac{y}{r_1^3} - \mu_2 \frac{y}{r_2^3}$$

$$r_1 = [(x - x_1)^2 + y^2]^{1/2} \qquad r_2 = [(x - x_2)^2 + y^2]^{1/2}$$

$$\mu_1 = Gm_1 \qquad\qquad\qquad \mu_2 = Gm_2$$

Here m_1, m_2 are the masses and $(x_1, 0)$, $(x_2, 0)$ are the
positions of the two massive bodies, and (x,y) is the position
of the satellite. These are Newton's equations which describe
the motion of the third body. The terms $-2\omega\, dy/dt$, $2\omega\, dx/dt$,
$\omega^2 x$, $\omega^2 y$ arise from the description of the motion using a
rotating frame of coordinates. The last two terms on the
right arise from the force of gravitational attraction.

The condition that there is a zero effective force acting
on the infinitesimal body in the rotating coordinate system
is that the pair of equations

$$\omega^2 x - \mu_1 \frac{(x - x_1)}{r_1^3} - \mu_2 \frac{(x - x_2)}{r_2^3} = 0$$

$$\omega^2 y - \mu_1 \frac{y}{r_1^3} - \mu_2 \frac{y}{r_2^3} = 0$$

have a solution. There are five positions $L_i = [\xi_i, \eta_i]$,
$i = 1, \ldots, 5$ lying in the xy plane (the plane of motion
of the two massive bodies) satisfying these equations. Three
of these lie along the x-axis and two of them lie at the
equilateral triangular points. If we insert a satellite into
one of these points and give it zero relative velocity (zero
velocity in the rotating frame), then by the equations of
motion it will experience no accelerations. It follows that
the satellite will remain at the point forever without the
application of any additional force. Thus the five particular
solutions found by Lagrange are given by

$$[x_i(t), y_i(t)] = [\xi_i, \eta_i] \quad i = 1, \ldots, 5$$

Approximate solutions of the above equations corresponding
to the three points lying along the x-axis can be obtained
by qualitative arguments. With a modest amount of algebraic
manipulation one can verify that the equilateral triangular
points also comprise solutions (see the exercises at the end
of this section).

In order to study the stability properties of these five solutions of Lagrange, we replace the above system of differential equations by an approximate system obtained by "linearizing." In order to study the stability properties of the triangular point $L_4 = [\xi_4, \eta_4]$, we introduce new coordinates which describe the deviation of the position of the satellite from the point L_4 and deviation of the velocity from zero:

$$\alpha = x - \xi_4 \qquad \beta = y - \eta_4$$

$$\gamma = \frac{d\alpha}{dt} \qquad \delta = \frac{d\beta}{dt}$$

and expand the forces up to first order using Taylor's theorem. Expressing these new equations as a system of four first order differential equations we have

$$\frac{d}{dt}\begin{bmatrix} \alpha \\ \beta \\ \gamma \\ \delta \end{bmatrix} = \begin{bmatrix} 0 & 0 & 1 & 0 \\ 0 & 0 & 0 & 1 \\ \frac{3}{4}\frac{G(m_1 + m_2)}{r_0^3} & \frac{3\sqrt{3}}{4}\frac{G(m_1 - m_2)}{r_0^3} & 0 & 2\omega \\ \frac{3\sqrt{3}}{4}\frac{G(m_1 - m_2)}{r_0^3} & \frac{9}{4}\frac{G(m_1 + m_2)}{r_0^3} & -2\omega & 0 \end{bmatrix}\begin{bmatrix} \alpha \\ \beta \\ \gamma \\ \delta \end{bmatrix}$$

where $r_0 = r_1 = r_2 = r_{12}$ at the triangular point. The stability properties of the linear system are determined by the coefficient matrix:

$$\begin{bmatrix} 0 & 0 & 1 & 0 \\ 0 & 0 & 0 & 1 \\ \dfrac{3}{4}\dfrac{G(m_1 + m_2)}{r_0^3} & \dfrac{3\sqrt{3}}{4}\dfrac{G(m_1 - m_2)}{r_0^3} & 0 & 2\omega \\ \dfrac{3\sqrt{3}}{4}\dfrac{G(m_1 - m_2)}{r_0^3} & \dfrac{9}{4}\dfrac{G(m_1 + m_2)}{r_0^3} & -2\omega & 0 \end{bmatrix}$$

If anyone cares to verify the linearized system, we have used the expression $\omega^2 = G(m_1 + m_2)/r_{12}^3$ in the development. (This formula arises in the solution of the two body problem and relates the angular frequency to the distance between the massive bodies in the case of circular orbits.)

The eigenvalues of this matrix are given by

$$\lambda_1 = -\lambda_2 = \sqrt{\frac{G(m_1 + m_2)}{2r_0^3}}\sqrt{-1 + \sqrt{1 - 27\frac{m_1 m_2}{(m_1 + m_2)^2}}}$$

$$\lambda_3 = -\lambda_4 = \sqrt{\frac{G(m_1 + m_2)}{2r_0^3}}\sqrt{-1 - \sqrt{1 - 27\frac{m_1 m_2}{(m_1 + m_2)^2}}}$$

The condition that these roots be pure imaginary is that

$1 - 27\dfrac{m_1 m_2}{(m_1 + m_2)^2} \geq 0$. Under these circumstances (see

Exercise 35 of Chap. 3) the linearized system is semi-stable; the trajectories of the satellite about the libration point are described by the sine and cosine functions and are periodic. If $1 - 27 (m_1 m_2)/(m_1 + m_2)^2 < 0$, then the system will be unstable since some of the roots will have a positive

real part. Introducing $\theta = m_1/m_2$ the condition for semi
stability can be written:

$$\theta^2 - 25\theta + 1 \geq 0$$

The roots of this equation are given by

$$\theta_{\pm} = \frac{25 \pm \sqrt{25^2 - 4}}{2}$$

Since $\theta_- \approx 0.0385$ it follows that $m_1/m_2 < 0.0385$ is the
condition that the solution be semi-stable.

A similar analysis of the points L_1, L_2, L_3 shows that
these libration points are all unstable.

Results regarding the stability properties of a system
obtained by this process of linearization are to be interpreted
carefully: If we find the linearized system to be either
stable or unstable, then the same holds for the "real" system.[†]
However semi-stability is a much more delicate matter. In
this case the qualitative behaviour of the original system
of differential equations is not determined by the qualitative
behaviour of the linearized system of equations. The semi-
stable solutions can be thought of as lying on the interface
between the stable and the unstable cases, and replacement
of the original system by the "linearized" system is simply
too crude an apprxoimation to allow adequate characterization
of this special case. The most convincing evidence regarding

[†] For a more detailed account of the relationship between the
stability properties of a differential equation and the
linearized differential equation, see Hahn (1967).

the semi-stability of the triangular points is, of course,
the existence of the minor planets at the triangular points
of the Sun-Jupiter system.

Example 2:   The Flying Eraser

An eraser normally has three different moments of
inertia:

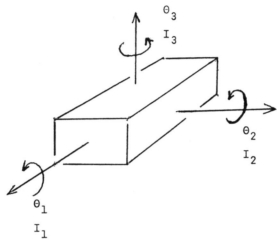

Here, I_1 is the smallest and I_3 is the largest moment of
inertia. The basic equations governing the motion of an
eraser flying through the air are those for a rotating rigid
body

$$\frac{dL}{dt} = \theta \times L$$

where L is the angular momentum vector

$$L = [I_1\theta_1, \ I_2\theta_2, \ I_3\theta_3]$$

and θ is the vector of angular velocities about the three axes
pictured above: $\theta = [\theta_1, \ \theta_2, \ \theta_3]$. The origin of the
coordinate system is at the centre of mass of the eraser and

the three coordinate directions are pointing in the directions
of the principle moments of the eraser. Thus, θ_1 is the
angular velocity about the first principle axis, and $I_1\theta_1$ is
the angular momentum about the same axis.

If an eraser is thrown through the air in such a fashion
that $\theta_1 \neq 0$, $\theta_2 = \theta_3 = 0$ initially, then we observe that all
of the rotational energy of the araser is maintained in the
motion about the θ_1 axis. That is, none of its rotational
energy will be transfered to rotation about either the θ_2 or
θ_3 axes. Similarly, if we throw the eraser in such a way
that $\theta_3 \neq 0$, $\theta_1 = \theta_2 = 0$ initially, then all of the rotational
energy will be maintained in the motion about the θ_3 axis.
However, if we throw the eraser so that initially $\theta_2 \neq 0$ and
$\theta_1 = \theta_3 = 0$, then in its flight through the air the eraser
will seem to flutter. The rotational energy will be transfered
from motion about the θ_2 axis to motion about the other two
principle axes θ_1 and θ_3.

We will give an argument that establishes that motion
about the θ_2 axis is unstable and indicates that motions
about the θ_1 and θ_3 axes are semi-stable. However a "flying
eraser" will always transfer its rotational energy into
rotations about either the θ_1 or θ_3 axis, the two axes
corresponding to the smallest and largest principle moments
of inertia. This may easily be verified by tossing an eraser
casually into the air several times taking care to impart
appropriate initial conditions to it.

In order to study the stability properties of the
motion about the three principle axes, we study the stability

properties of the solution of the equation of motion obtained using the three sets of initial data ($\theta_i(0) = \omega_0$, $\theta_j(0) = 0$, $j \neq i$, $i = 1, 2, 3$). The stability properties of these solutions are studied by "linearizing" the equation of motion as we did above. We consider, in detail, the case where $\theta_1(0) = \omega_0$, $\theta_2(0) = \theta_3(0) = 0$. Rewriting the equation of motion as

$$
\begin{bmatrix}
I_1 \frac{d}{dt} \theta_1 \\[2mm]
I_2 \frac{d}{dt} \theta_2 \\[2mm]
I_3 \frac{d}{dt} \theta_3
\end{bmatrix}
=
\begin{bmatrix}
(I_3 - I_2)\theta_2\theta_3 \\[2mm]
(I_1 - I_3)\theta_1\theta_3 \\[2mm]
(I_2 - I_1)\theta_1\theta_2
\end{bmatrix}
$$

letting $\theta_i(t) = \theta_i(0) + \Delta\theta_i(t)$, $i = 1, 2, 3$, and dropping terms that are second order in the small quantities $\Delta\theta_i$ yields

$$
\frac{d}{dt}
\begin{bmatrix}
\Delta\theta_1 \\[2mm]
\Delta\theta_2 \\[2mm]
\Delta\theta_3
\end{bmatrix}
=
\begin{bmatrix}
0 & 0 & 0 \\[2mm]
0 & 0 & \dfrac{I_1 - I_3}{I_2}\omega_0 \\[2mm]
0 & \dfrac{I_2 - I_1}{I_3}\omega_0 & 0
\end{bmatrix}
\begin{bmatrix}
\Delta\theta_1 \\[2mm]
\Delta\theta_2 \\[2mm]
\Delta\theta_3
\end{bmatrix}
$$

The eigenvalues of the matrix

$$
\begin{bmatrix}
0 & 0 & 0 \\[2mm]
0 & 0 & \dfrac{I_1 - I_3}{I_2}\omega_0 \\[2mm]
0 & \dfrac{I_2 - I_1}{I_3}\omega_0 & 0
\end{bmatrix}
$$

are given by 0,

$$\lambda_{\pm} = \pm \sqrt{\frac{(I_1 - I_3)(I_2 - I_1)}{I_2 I_3} \omega_0^2}$$

and since $I_1 < I_2 < I_3$, the eigenvalues $\lambda_{\pm}$ are pure imaginary.
Thus this linearized system is semi-stable.

A similar computation can be carried out for motion
about the θ_3 axis. In this case the eigenvalues of the
corresponding matrix are given by 0,

$$\lambda_{\pm} = \pm \sqrt{\frac{(I_3 - I_1)(I_2 - I_3)}{I_1 I_2} \omega_0^2}$$

Again the eigenvalues $\lambda_{\pm}$ are pure imaginary and the linearized
system is semi-stable. However if we linearize about the
point $\theta_2 = \omega_0$, $\theta_1 = \theta_3 = 0$, the eigenvalues in question are
given by 0,

$$\lambda_{\pm} = \pm \sqrt{\frac{(I_2 - I_1)(I_3 - I_2)}{I_1 I_3} \omega_0^2}$$

and all of the eigenvalues are real and one is positive.
Thus the linearized system is unstable in this case.

Regarding the stability of the various types of motion
considered in the original problem, our results show that
motion about the θ_2 axis is unstable but only indicate the
motion about the other two axes is semi-stable. Of course,
the most concrete piece of evidence regarding the stability
properties of the angular motion can be obtained by throwing
an eraser through the air, giving it various types of

rotational energy, and then observing what happens.

Exercises

51. Show using qualitative arguments that there are three
     solutions of the equations:

$$\omega^2 x - \mu_1 \frac{x - x_1}{r_1^3} - \mu_2 \frac{x - x_2}{r_2^3} = 0$$

$$\omega^2 y - \mu_1 \frac{y}{r_1^3} - \mu_2 \frac{y}{r_2^3} = 0$$

     lying along the x-axis.

52. By solving the equation of motion for the two body
     problem and assuming circular motion one finds that the
     angular frequency, the distance between the two masses
     r_{12} and the two masses are related by the formula

$$\omega^2 = G(m_1 + m_2)/r_{12}^3$$

     where G is the gravitational constant. Using the
     formula, show that the resultant forces in the rotating
     frame are zero at the two equilateral triangular points.

CHAPTER 4

LOCATION AND ESTIMATION OF EIGENVALUES

Very often a rough approximation of the eigenvalues of
a matrix is all that's necessary in a particular problem.
For example, just knowing that $|M| \leq \max_j \sum_i |m_{ij}|$ enables us
to conclude that if $A = \begin{bmatrix} 0.8 & 0.3 \\ 0.1 & 0.5 \end{bmatrix}$, then $|A| \leq 0.9 < 1$, and
hence that $\sum_{n=0}^{\infty} \begin{bmatrix} 0.8 & 0.3 \\ 0.1 & 0.5 \end{bmatrix}^n = (I - A)^{-1} = \begin{bmatrix} 0.2 & -0.3 \\ -0.1 & 0.5 \end{bmatrix}^{-1} =$
$\frac{10}{7} \begin{bmatrix} 5 & 3 \\ 1 & 2 \end{bmatrix}$ without computing any eigenvalues at all.

In our discussion of stability we saw that all we need
to know to conclude that A is a stability matrix is that
$Re(\lambda) < 0$ for all eigenvalues λ, so if we knew, for example,
that every eigenvalue of a matrix A were in a disc of radius
2 about the point -3 + i, then we could conclude that A is
a stability matrix even though this estimate of the eigenvalues
of A is very crude.

We shall begin our investigation of these estimates in
Sec. 4. In preparation for that discussion we shall review
orthogonality in $\underline{C}^k$ in the next three sections.

1. AN INNER PRODUCT IN $\underline{C}^k$

It is customary to define a scalar-valued function (x,y) of
two vector variables x, y by

$$(x, y) = y^* x \quad \text{for all } x, y \in \underline{C}^k$$

It is easy to verify that this function (called the *standard
inner product*) has the following properties for every x, y,
$z \in \underline{C}^k$ and every α, $\beta \in \underline{C}$:

1. $(\alpha x + \beta y, z) = \alpha(x, z) + \beta(y, z)$, i.e., the inner[†]
    product is linear in the first variable;

2. $(y, x) = \overline{(x,y)}$, i.e., the inner product is "conjugate
    symmetric;" and hence

3. $(x, \alpha y + \beta z) = \bar{\alpha}(x, y) + \bar{\beta}(x, z)$, i.e., the inner
    product is "conjugate-linear" in the second variable.

The inner product defines a function on $\underline{C}^k$ (called the
Euclidean norm) by

$$|x| = \sqrt{(x,x)} \quad \text{for all } x \in \underline{C}^k$$

We saw in Chap. 3, Sec. 10 that (x, x) is real for all $x \in \underline{C}^k$,
in fact that $(x, x) > 0$ for all $x \neq 0$. Thus $|x| = 0$ iff
$x = 0$. Here are two other basic properties of the Euclidean
norm: For all x, $y \in \underline{C}^k$ and all $\alpha \in \underline{C}$:

4. $|x + y| \leq |x| + |y|$ (the triangle inequality)

5. $|\alpha x| = |\alpha||x|$ (homogeneity)

A vector u is said to be a *unit vector* iff $|u| = 1$.

[†]
  Hereafter we'll call "the standard inner product" the
  "inner product" for short.

If $(x, x) = 0$ we say that x *is orthogonal to* y and we may write $x \perp y$. If S is any subset of $\underline{C}^k$ we say that y *is orthogonal to* S and write $y \perp S$ iff $y \perp s$ for all $s \in S$. The *orthogonal complement* $S^\perp$ of S is the set of all vectors orthogonal to S, that is, $S^\perp = \{y \in \underline{C}^k : y \perp S\}$.

Exercises

1. Show that $S^\perp$ is a subspace of $\underline{C}^k$ for any subset S of $\underline{C}^k$.
2. Prove "Pythagoras' theorem":

If x, y, z are any points in $\underline{C}^k$ and $x - y \perp y - z$, then $|x - y|^2 + |y - z|^2 = |x - z|^2$.

You have, of course, noticed that the inner product reduced to the old familiar dot product when restricted to real vectors, that is, $(x, y) = y^{tr}x = y \cdot x = x \cdot y$ when x, $y \in \underline{R}^k$.

2. UNITARY MATRICES AND ORTHONORMAL BASES

Which $k \times k$ complex matrices preserve the inner product (and hence the norm, orthogonality, etc.)? Suppose $(My, Mx) = (y, x)$ for all x, $y \in \underline{C}^k$. Then $x^* M^* My = x * y$ for all x, y; in particular, for $x = e_j$ and $y = e_\ell$, where e_s is the sth column of I. Therefore

$$e_j^* M^* M e_\ell = \begin{cases} 1 & \text{if } j = \ell \\ 0 & \text{if } j \neq \ell \end{cases}$$

On the other hand, $e_j^* N e_\ell$ is the $j\ell^{th}$ entry in N, so $M^* M = I$.

Conversely, if $M^*M = I$, then clearly $(Mx)^*(My) = x^*y$ for all

x, y ϵ $\underline{C}^k$; so the requirement that M preserve the inner

product is equivalent to the requirement that $M^* = M^{-1}$. We

call matrices with this property *unitary* matrices. If they

happen to be real as well, they are called *orthogonal* matrices.

It's easy to see if a given matrix M is unitary, because

we need only check to see if $M^*M = I$, i.e., if the columns of

M are pairwise orthogonal unit vectors. For example,

$1/\sqrt{2} \begin{bmatrix} 1 & i \\ -i & -1 \end{bmatrix}$ is unitary and $1/\sqrt{2} \begin{bmatrix} 1 & 1 \\ 1 & -1 \end{bmatrix}$ is orthogonal. There-

fore the columns of any unitary matrix are an example of an

"orthonormal" set:

A set $\{u^{(1)}, u^{(2)}, \ldots, u^{(s)}\}$ of vectors in $\underline{C}^k$ is

orthonormal iff $u^{(\ell)} \perp u^{(j)}$ when $\ell \neq j$ and $|u^{(j)}| = 1$ for

all ℓ, $j \leq s$. Such sets are always linearly independent for

if

$$\sum_{j=1}^{s} \alpha_j u^{(j)} = 0$$

then

$$0 = \left(\left[\sum_{j=1}^{s} \alpha_j u^{(j)} \right], u^{(\ell)} \right)$$

$$= \sum_{j=1}^{s} \alpha_j (u^{(j)}, u^{(\ell)})$$

$$= \alpha_\ell \text{ for } \ell = 1, 2, \ldots, s$$

Thus any orthonormal set with k elements (for example, the

columns of a unitary matrix) is a *basis* for $\underline{C}^k$ -- called, of

course, an *orthonormal basis*. These are useful because the formula for finding the coordinates of a vector (for "resolving" the vector) with respect to such a basis is so simple: Suppose α_1, α_2, ..., α_k are the coordinates of v with respect to the orthonormal basis $\{u^{(1)}, \ldots, u^{(k)}\}$, then $v = \sum_{j=1}^{s} \alpha_j u^{(j)}$. How are the α's related to v and the u's? It's just $\alpha_j = (v, u^{(j)})$ for each $1 \leq j \leq k$.

We observed that the columns of any unitary matrix must form an orthonormal basis for $\underline{C}^k$. Conversely, if $u^{(1)}$, ..., $u^{(k)}$ are an orthonormal basis, then the matrix U whose j^{th} column is $u^{(j)}$ is unitary.

Theorem (Gram-Schmidt): Every unit vector in $\underline{C}^k$ is part of some orthonormal basis for $\underline{C}^k$.

Proof: Suppose u is a unit vector in $\underline{C}^k$. If k > 1, let $u^{(1)} = u$ and S be the subspace spanned by $u^{(1)}$. Choose $v \notin S$ and let $w = v - (v, u^{(1)})u^{(1)}$. Then $u^{(1)} \perp w$ and $w \neq 0$, and hence if we put $u^{(2)} = w/|w|$, then $\{u^{(1)}, u^{(2)}\}$ is an orthonormal set.

If k > 2, let S be the subspace spanned by $\{u^{(1)}, u^{(2)}\}$. Choose $v \notin S$ and let $w = v - (v, u^{(1)})u^{(1)} - (v, u^{(2)})u^{(2)}$. Then $u^{(j)} \perp w$ for j = 1, 2 and $w \neq 0$, and hence if we put $u^{(3)} = w/|w|$ then $\{u^{(1)}, u^{(2)}, u^{(3)}\}$ is an orthonormal set.

If k > 3, let S be the subspace spanned by $\{u^{(1)}, u^{(2)}, u^{(3)}\}$ ... continuing similarly ... then $\{u^{(1)}, u^{(2)}, \ldots, u^{(k-1)}\}$ is an orthonormal set.

Let S be the subspace spanned by $\{u^{(1)}, \ldots, u^{(k-1)}\}$.

Choose $v \notin S$ and let $w = v - \sum_{j=1}^{k-1} (v, u^{(j)})u^{(j)}$. Then

$u^{(j)} \perp w$ for $j = 1, 2, \ldots, k-1$ and $w \neq 0$, and hence if we

put $u^{(k)} = w/|w|$, then $\{u^{(1)}, u^{(2)}, \ldots, u^{(k)}\}$ is an

orthonormal set of k vectors and is therefore an orthonormal

basis. []

Exercise

3. If S is a subspace of $\underline{C}^k$, show that $S = S\perp\perp$.

3. SCHUR'S THEOREM

A matrix T is said to be *upper triangular* iff $t_{ij} = 0$, whenever

$i > j$; thus for example,

$$T = \begin{bmatrix} 1 & 2 & 3 \\ 0 & 0 & 2 \\ 0 & 0 & -1 \end{bmatrix}$$

is upper triangular. Every Jordan matrix is upper triangular.

A matrix is said to be *lower triangular* if its transpose

is upper triangular.

Diagonal matrices are the only matrices which are both

upper and lower triangular.

Questions concerning the inner product usually cannot be

handled by an examination of the Jordan form because the

matrix P that puts A into the Jordan form J ($A = PJP^{-1}$) may

not preserve the inner product. We shall soon see, however,

that every matrix is unitarily similar to an upper triangular

matrix (that is, $A = UTU^{-1}$, where U is unitary and T is upper

triangular). Since unitary matrices preserve the inner
product, if we can settle our problem for triangular matrices,
we can settle it for any matrix. Moreover triangular matrices
are almost as easy to deal with as Jordan matrices for many
purposes.

Theorem (Schur): Every matrix is unitarily similar to
some upper triangular matrix. (That is, if A is any k × k
complex matrix, then there exists a unitary U and upper
triangular T such that $A = UTU^*$.)

Proof (by induction on k): If k = 1 we're through.
Suppose the proposition is true for all $\ell \times \ell$ matrices with
$\ell < k$.

$Aw = \lambda w$ for some $0 \neq w \in \underline{C}^k$ and some $\lambda \in \underline{C}$. Let $u^{(1)} = |w|^{-1}w$ then, by Gram-Schmidt, $u^{(1)}$ is part of an orthonormal
basis for $\underline{C}^k$: $\{u^{(1)}, u^{(2)}, \ldots, u^{(k)}\}$. If we let W be the
matrix whose j^{th} column is $u^{(j)}$, then, as we observed above,
W must be unitary. Moreover, if we let $B = W^{-1}AW$, then the
first column of B is $[\lambda, 0, 0, \ldots, 0]^{tr}$ because if e_1 is the
first column of I, then the first column of B is Be_1 and we
have $Be_1 = W^{-1}AWe_1 = W^{-1}Au^{(1)} = W^{-1}\lambda u^{(1)} = \lambda W^{-1}We_1 = \lambda e_1$.
If we let $c_{ij} = b_{i+1,j+1}$ for $1 \leq i, j \leq k - 1$, we can write

$$B = \begin{bmatrix} \lambda & b_{12} & b_{13} & \cdots & b_{1k} \\ 0 & & & & \\ 0 & & & & \\ \vdots & & & C & \\ 0 & & & & \end{bmatrix}$$

Now $C = VSV^*$ for some upper triangular (k - 1) × (k - 1)
matrix S and some (k - 1) × (k - 1) unitary matrix V.

Therefore

$$
W^* AW = \begin{bmatrix} \lambda & b_{12} & b_{13} & \cdots & b_{1k} \\ 0 \\ 0 & & VSV^* \\ \vdots \\ 0 \end{bmatrix}
$$

$$
= (1 \oplus V) \begin{bmatrix} \lambda & b'_{12} & b'_{13} & b'_{13} & \cdots & b'_{1k} \\ 0 & s_{11} & s_{12} & s_{23} & \cdots & s_{1k-1} \\ 0 & 0 & s_{22} & s_{23} & \cdots & s_{2k-1} \\ 0 & 0 & 0 & s_{33} & \cdots & s_{3k-1} \\ \vdots & \vdots & \vdots & \vdots & \cdots & \vdots \\ 0 & 0 & 0 & 0 & & s_{k-1,k-1} \end{bmatrix} (1 \oplus V)^*
$$

where $b'_{1j} = \sum_{\ell=2}^{k} b_{1\ell} v_{\ell j}$. Consequently, $A = UTU^*$, where $U = W(1 \oplus V)$, and T is the upper triangular matrix

$$
\begin{bmatrix} \lambda & b'_{12} & b'_{13} & \cdots & b'_{1k} \\ 0 \\ 0 & & S \\ \vdots \\ 0 \end{bmatrix}
$$

U is unitary because $(1 \oplus V)$ and W are unitary and the product of unitary matrices is unitary. []

Corollary 1: If H is Hermitian, then H is unitarily similar to a real diagonal matrix.

Proof:

$$
H = UTU^*
$$

where U is unitary and T is upper triangular.

$$T = U^*HU \quad \text{as } U^* = U^{-1}$$
$$T^* = U^*H^*U \quad \text{as } (MN)^* = N^*M^*$$
$$= U^*HU \quad \text{because } H \text{ is Hermitian}$$
$$= T$$

But T^* is a lower triangular matrix (being the transpose of an upper triangular matrix), therefore, T is both upper and lower triangular and is therefore diagonal. Since $t_{ii}^* = t_{ii}$, all the entries in T are real. []

  Corollary 2. If H is Hermitian, then H is positive definite iff all its eigenvalues are positive.

Exercises

4. Show that every real matrix, all of whose eigenvalues are real, is orthogonally similar to some real upper triangular matrix.

5. Show that every real symmetric matrix is orthogonally similar to a real diagonal matrix.

6. Prove corollary 2.

7. a. If A is unitarily diagonable, that is, $A = UDU^*$, where D is diagonal and U is unitary, show that $AA^* = A^*A$. (*Note:* A is said to be *normal* when $AA^* = A^*A$.)

   b. Prove the converse of the proposition established in (a). [*Hint:* $A = U^*TU$. Show T is normal then see what can be said about normal triangular matrices by multiplying out T^*T and TT^* and comparing entries (do this for $k = 3$ to get the idea).]

8. Show that in Shur's triangularization theorem, the
    eigenvalues of A can be made to appear in any prescribed
    order on the main diagonal.

4. EIGENVALUES OF HERMITIAN MATRICES

Suppose H is Hermitian and $H = UDU^*$, where D is diagonal and
U is unitary as in corollary 1. $D = \lambda_1 \oplus \lambda_2 \oplus \ldots \oplus \lambda_k$,
where the λ_i are the eigenvalues of H. We may assume, without
loss of generality, that $\lambda_1 \geq \lambda_2 \geq \ldots \geq \lambda_k$ (see Exercise 9).
Let us set $y = U^*x$, then for every $x \in \underline{C}^k$,

$$x^*Hx = x^*UDU^*x$$

$$= y^*Dy$$

$$= \sum_{j=1}^{k} \lambda_j |y_j|^2$$

It follows that $\lambda_k y^*y \leq x^*Hx \leq \lambda_1 y^*y$. But $y^*y = x^*x$ because
U being unitary preserves the inner product. Therefore

$$\lambda_k \leq \frac{x^*Hx}{x^*x} \leq \lambda_1 \quad \text{for all } 0 \neq x \in \underline{C}^k \qquad (1)$$

The (real) number x^*Hx/x^*x is called the *Rayleigh quotient*
and it is denoted by $q_H(x)$. Notice that if $q_H(x) = \alpha$, then
there is unit vector $u = x/|x|$ such that $q_h(u) = \alpha$; so the
set of values taken on by q_H is the set of values $\{q_H(u) :$
$|u| = 1\}$. Since $q_H(w) = \lambda_1$ and $q_H(v) = \lambda_k$ when w and v are
eigenvectors for λ_1 and λ_k we see that:

    Theorem 1: If λ_1 is the largest eigenvalue of a
Hermitian matrix and λ_k is its smallest eigenvalue, then

$$\lambda_1 = \max_{x \neq 0} q_H(x) = \max_{|u|=1} q_H(u)$$

$$\lambda_k = \min_{x \neq 0} q_H(x) = \min_{|u|=1} q_H(u)$$

Exercise

9. Suppose $H = VEV^*$, where E is diagonal and V is unitary.
    Show that there exists a unitary matrix P such that
    $P^{-1}EP = \lambda_1 \oplus \lambda_2 \oplus \ldots \oplus \lambda_k$, where $\lambda_j \geq \lambda_{j+1}$ ($1 \leq j \leq$ k-1)
    and hence that $H = U(\lambda_1 \oplus \lambda_2 \oplus \ldots \oplus \lambda_k)U^*$ for some
    unitary U.

The following theorem is a generalization of theorem 1
which explains in a variational way how one goes from a
determination of a unit eigenvector for the jth largest
eigenvalue to a unit vector for the $j + 1^{st}$ largest eigenvalue.

Theorem 2. Suppose $\lambda_1 \geq \lambda_2 \geq \ldots \geq \lambda_k$ are the eigenvalues
of a Hermitian matrix H and $\{u^{(1)}, u^{(2)}, \ldots, u^{(k)}\}$ is an
orthonormal set of eigenvectors for the λ_j ($Hu^{(j)} = \lambda_j u^{(j)}$).
If $S_0 = \{0\}$ and $S_j = \{u^{(1)}, u^{(2)}, \ldots, u^{(j)}\}$ for $1 \leq j \leq$ k,
then

$$\lambda_j = \text{lub}\{q_H(x) : 0 \neq x \in S_{j-1}^{\perp}\} \quad \text{for some } 1 \leq j \leq k$$

Proof: Let $x \in \underline{C}^k$. $x = \sum_{\ell=1}^{k} \xi_\ell u^{(\ell)}$, where $\xi_\ell = (x, u^{(\ell)})$.

If $x \in S_{j-1}^{\perp}$, then $\xi_1 = \xi_2 = \ldots = \xi_{j-1} = 0$, and hence $x = \sum_{\ell=j}^{k} \xi_\ell u^{(\ell)}$. Therefore $Hx = \sum_{\ell=j}^{k} \xi_\ell \lambda_\ell u^{(\ell)}$, and hence

$$x^* Hx = \sum_{t=j}^{k} \sum_{\ell=j}^{k} \xi_t^* \lambda_\ell \xi_\ell (u^{(t)*} u^{(\ell)})$$

$$= \sum_{\ell=j}^{k} \lambda_\ell |\xi_\ell|^2 \leq \lambda_j \sum_{\ell=j}^{k} |\xi_\ell|^2$$

Now $|x|^2 = \sum_{\ell=1}^{k} |\xi_\ell|^2$; so $x^* Hx \leq \lambda_j |x|^2$, and hence

$$q_H(x) \leq \lambda_j \quad \text{for all } 0 \neq x \in S_{j-1}^{\perp}$$

Since $q_H(u^{(j)}) = \lambda_j$ and $0 \neq u^{(j)} \in S_{j-1}^{\perp}$, we have shown that

$$\lambda_j = \text{lub}\{q_H(x) : 0 \neq x \in S_{j-1}^{\perp}\}$$

5. BOUNDS FOR THE EIGENVALUES OF ARBITRARY MATRICES: ESTIMATING THE SPECTRAL RADIUS

Suppose A is a k × k complex matrix. A k-tuple $(\lambda_1, \lambda_2, \ldots, \lambda_k)$ consisting of the eigenvalues of A each occurring as often as its multiplicity is sometimes called a *spectrum* of A. It is customary, if slightly inaccurate, to call any such k-tuple "the" spectrum of A. For example:

matrix	spectrum
$\begin{bmatrix} 1 & 2 \\ 1 & 2 \end{bmatrix}$	(0, 3) also (3, 0)
$\begin{bmatrix} 1 & 2 & 0 \\ 0 & 1 & 2 \\ 0 & 0 & 3 \end{bmatrix}$	(1, 3, 1), (1, 1, 3), and (3, 1, 1)
$\begin{bmatrix} 1 & 0 & 2 \\ 0 & 1 & 0 \\ 0 & 0 & 1 \end{bmatrix}$	(1, 1, 1) only

Exercises

10. The diagonal $(t_{11}, t_{22}, \ldots, t_{kk})$ of an upper triangular matrix T is a spectrum of T.

11. Spectrum is a similarity invariant.

12. If M is any $k \times k$ matrix then

$$\text{trace}(M^*M) = \sum_{i=1}^{k} \sum_{j=1}^{k} |m_{ij}|^2$$

If we think of A as a vector in $\underline{C}^{k^2}$ and its spectrum as a vector in $\underline{C}^k$, then it turns out that the Euclidean norm of the spectrum doesn't exceed the Euclidean norm of A. This is essentially the content of Schur's inequality.

Schur's Inequality: If A is a complex $k \times k$ matrix and $(\lambda_1, \lambda_2, \ldots, \lambda_k)$ is its spectrum, then

$$\sum_{i=1}^{k} |\lambda_i|^2 \leq \sum_{i=1}^{k} \sum_{j=1}^{k} |a_{ij}|^2$$

Proof: $A = UTU^*$, where U is unitary and T is upper triangular. Exercises 10 and 11 imply that $(t_{11}, t_{22}, \ldots, t_{kk})$ is a spectrum of A; so

$$\sum_{i=1}^{k} |\lambda_i|^2 = \sum_{j=1}^{k} |t_{jj}|^2$$

$$\leq \sum_{i=1}^{k} \sum_{j=1}^{k} |t_{ij}|^2 = \text{trace}(T^*T) \quad \text{(by Exercise 12)}$$

Now $T^*T = UA^*AU^{-1}$ and trace is a similarity invariant, so

$$\text{trace}(T^*T) = \text{trace}(A^*A)$$

$$= \sum_{i=1}^{k} \sum_{j=1}^{k} |a_{ij}|^2 \quad \text{(by Exercise 12)}$$

Therefore

$$\sum_{i=1}^{k} |\lambda_i|^2 \leq \sum_{i=1}^{k} \sum_{j=1}^{k} |a_{ij}|^2$$

Exercise

13. Show that equality holds in Schur's inequality iff
     $AA^* = A^*A$. (Suggestion: See Exercise 7.)

     For every k × k complex matrix A, let $\hat{A} = \frac{1}{2}(A + A^*)$ and
$\tilde{A} = \frac{i}{2}(A^* - A)$.

Exercises

14. Show that $|A| \leq \sqrt{\text{trace}(A^*A)}$.
15. Let A be any k × k complex matrix
     a. Show that $\hat{A}$ and $\tilde{A}$ are Hermitian.
     b. If $\hat{\lambda}_1$ is the largest eigenvalue of $\hat{A}$
         $\tilde{\lambda}_1$ is the largest eigenvalue of $\tilde{A}$
         $\hat{\lambda}_k$ is the smallest eigenvalue of $\hat{A}$
         $\tilde{\lambda}_k$ is the smallest eigenvalue of $\tilde{A}$
         and λ is any eigenvalue of A, show that $\hat{\lambda}_k \leq \text{Re}(\lambda) \leq$
         $\hat{\lambda}_1$ and $\tilde{\lambda}_k \leq \text{Im}(\lambda) \leq \tilde{\lambda}_1$.
     c. Show that $|\text{Re}(\lambda)| \leq |\hat{A}|$ and $|\text{Im}(\lambda)| \leq |\tilde{A}|$.
     d. Show that $|A| \leq \sqrt{|\hat{A}|^2 + |\tilde{A}|^2}$.

16.   Let A by any k × k complex matrix.

   a.   Show that A^*A is Hermitian and no eigenvalue of A^*A is negative. If $(\alpha_1, \alpha_2, \ldots, \alpha_k)$ is a spectrum of A^*A, then the numbers $\sqrt{\alpha_1}, \sqrt{\alpha_2}, \ldots, \sqrt{\alpha_k}$ are called the *singular values* of A.

   b.   If $\sqrt{\alpha_1}$ is the largest and $\sqrt{\alpha_k}$ is the smallest singular value of A, show that $\sqrt{\alpha_k} \le |\lambda| \le \sqrt{\alpha_1}$ for every eigenvalue λ of A.

   c.   Show that $|A| \le \sqrt{|A^*A|}$.

   d.   Use (c) and an estimate of $|A^{tr}A|$ based on Chap. 2, Exercise 10 to estimate the spectral radius of

$$A = \begin{bmatrix} 1 & 2 & -1 \\ -2 & 1 & 2 \\ 0 & 2 & 1 \end{bmatrix}.$$ Compare with the estimate of $|A|$

obtained by using Chap. 2, Exercise 10 directly.

17.   a.   If $B = \begin{bmatrix} 1 & 2 & -1 \\ -2 & 1 & 1 \\ 0 & 2 & 1 \end{bmatrix}$ estimate $|B|$ using first Exercise

10 of Chap. 2 then Exercise 14, then 15(d) and finally the method of 16(d).   (Compare.)

   b.   Repeat (a) with the matrix A of 16(d).

18.   If $A = \begin{bmatrix} 1 & -1 & 2 & 3 & -2 \\ 3 & 2 & 3 & -2 & 2 \\ 4 & 3 & 0 & 1 & 2 \\ -1 & 2 & 3 & 1 & -1 \\ 4 & -2 & 2 & 1 & 1 \end{bmatrix}$, show that $|A| < 10$.

19.   If $A = \begin{bmatrix} 0.1 & 0.2 & 0.4 & 0.3 \\ 0.2 & 0.3 & 0.8 & 0 \\ 0.4 & 0.8 & 0.1 & 0 \\ 0.3 & 0 & 0 & 0.1 \end{bmatrix}$ does $\sum_{n=1}^{\infty} \frac{1}{n} A^n$ converge?

6. REGIONS CONTAINING THE EIGENVALUES OF A MATRIX: GERSGORIN'S THEOREM

Gerŝgorin's theorem has proven to be a very useful way of locating the eigenvalues of a matrix approximately. As you will see shortly the proof of this theorem takes just two sentences, but the remarkable thing here is that so useful a theorem and one with such a simple proof had to wait until as recently as 1931 to be discovered.

If A is an arbitrary k × k complex matrix, let

$$D_j(A) = \{z \in C : |z - a_{jj}| \leq \sum_{\ell \neq j} |a_{j\ell}|\} \quad \text{for } j = 1, 2, \ldots, k$$

These sets $D_j(A)$ are discs in the complex plane. They are called the *Gerŝgorin discs* of A. For example, if

$$A = \begin{bmatrix} 3 & 2 & 2 & -4 \\ 2 & 3 & 2 & -1 \\ 1 & 1 & 2 & -1 \\ 2 & 2 & 2 & -1 \end{bmatrix}$$

then D_1 is the disc centered at 3 of radius 8, D_2 is the disc centered at 3 of radius 5, D_3 is the disc centered at 2 of radius 3, and D_{II} is the disc centered at -1 of radius 6.

Gerŝgorin's Theorem: Each eigenvalue of A lies in some Gerŝgorin disc of A.

Proof: Let λ be any eigenvalue of A, then for some x ≠ 0 we have Ax = λx. Suppose j is an index such that $|x_j| \geq |x_\ell|$ for all $\ell = 1, 2, \ldots, k$, then $|x_j| \neq 0$ and

$$\lambda x_j = \sum_{\ell=1}^{k} a_{j\ell} x_\ell = a_{jj} x_j + \sum_{\ell \neq j} a_{j\ell} x_\ell$$

therefore

$$|\lambda - a_{jj}||x_j| \le \sum_{\ell \ne j} |a_{j\ell}||x_\ell| \le \left(\sum_{\ell \ne j} |a_{j\ell}| \; |x_j|\right)$$

and hence $|\lambda - a_{jj}| \le \sum_{\ell \ne j} |a_{j\ell}|$; consequently, $\lambda \in D_j(A)$. []

Exercises

20. Derive the result of Exercise 10 in Chap. 2 from Gerŝgorin's theorem.

21. Use Gerŝgorin's theorem to prove that

$$\begin{bmatrix} 4 & 1 & -2 & 0 \\ -2 & 5 & 1 & 1 \\ 3 & 1 & 6 & -1 \\ 1 & -1 & 2 & 5 \end{bmatrix}$$

    is invertible.

22. Let $\gamma(A) = \min_j \{|a_{jj}| - \sum_{\ell \ne j} |a_{j\ell}|\}$. Show that $|\lambda| \ge \gamma(A)$ for all eigenvalues λ of any matrix A.

23. Show that $|\det A| \ge \gamma^k(A)$ if $\gamma(A) > 0$ and A is any $k \times k$ complex matrix.

24. Show that $\begin{bmatrix} 8 & 2 & 0 & 2 \\ 0 & 5 & 1 & 1 \\ 1 & -1 & 8 & -2 \\ 1 & 1 & 1 & 6 \end{bmatrix}$ is not a stability matrix.

25. If A is a singular $k \times k$ complex matrix, show that for some $i \le k$, $|a_{ii}| \le \sum_{j \ne i} |a_{ij}|$.

26. If A is a real $k \times k$ matrix and

$$a_{ii} < -\left(\sum_{j \ne i} |a_{ij}|\right) \text{ for } i = 1, 2, \ldots, k$$

    show that A is a stability matrix.

27. Prove that if A is a stochastic matrix and $\omega = \min_{i} a_{ii}$

then for every eigenvalue λ of A we have

$$|\lambda - \omega| \leq 1 - \omega$$

While it is not true that every Gerŝgorin disc contains

an eigenvalue of A [e.g., if $A = \begin{bmatrix} -4 & -10 \\ 1 & 6 \end{bmatrix}$, then $D_2(A)$

contains neither eigenvalue of A], it is true that a union

of discs which is disjoint from all the other discs must

contain an eigenvalue of A:

Theorem. (Gerŝgorin, 1931): Suppose $(i_1, i_2, \ldots, i_k)$

is a permutation of $\{1, 2, \ldots, k\}$, $S = \bigcup_{j=1}^{s} D_{i_j}(A)$, and

$T = \bigcup_{j=s+1}^{k} D_{i_j}(A)$. If $S \cap T = \emptyset$, then S contains some eigenvalue

of A.

Before proceeding to the proof we have to present the

following:

Lemma: If $\lim_{n \to \infty} A_n = A$, then λ is an eigenvalue of A if,

and only if, there exists a sequence of eigenvalues λ_n of

A_n such that $\lim_{n \to \infty} \lambda_n = \lambda$.

Proof: Let $c_n(\tau)$ and $c(\tau)$ denote the characteristic

polynomials of A_n and A. Because the determinant of M is a

polynomial in the k^2 variables m_{ij}, it is a continuous function,

and hence $\lim_{n \to \infty} \det(\tau I - A_n) = \det(\tau I - A)$. Therefore:

a. $\lim_{n \to \infty} c_n(\tau) = c(\tau)$ for all τ

We also know that

b. $\quad c_n(\tau) = \prod_{j=1}^{k} (\tau - \lambda_j^{(n)})$ where $(\lambda_1^{(n)}, \ldots, \lambda_k^{(n)})$

is a spectrum of A_n

$\qquad c(\tau) = \prod_{j=1}^{k} (\tau - \lambda_j)$ $\qquad$ where $(\lambda_1, \ldots, \lambda_k)$ is a

spectrum of A

Suppose $\quad$ is an eigenvalue of A. According to (a),

$\lim_{n \to \infty} c_n(\lambda) = 0$. Therefore, given any positive integer m,

there exists a positive integer N_m such that

$$|c_n(\lambda)| < (\frac{1}{m})^k \quad \text{for all } n \geq N_m$$

Therefore applying (b) we see that if $n \geq N_m$, there is some

index $j_{n,m}$ between 1 and k for which $|\lambda - \lambda_j^{(n)}| < \frac{1}{m}$. Without

loss of generality, we may assume that $N_m < N_{m+1}$ for all

$m \geq 1$. Therefore if $n \geq N_1$, then there exists a unique m

such that $N_m \leq n < N_{m+1}$, and we define $\lambda_n = \lambda_{j_{n,m}}^{(n)}$; if $n < N_1$,

we let $\lambda_n = \lambda_1^{(n)}$. Since λ_n is then an eigenvalue of A_n for

all n, it remains to prove that $\lim_{n \to \infty} \lambda_n = \lambda$. Let ε be any

positive number. Choose an integer p such that $0 < \frac{1}{p} \leq \varepsilon$.

If $n \geq N_p$, we then have $N_p \leq N_m \leq n < N_{m+1}$ for some $m \geq p$,

and hence

$$|\lambda - \lambda_n| = |\lambda - \lambda_{j_{n,m}}^{(n)}| < \frac{1}{m} \leq \frac{1}{p} \leq \varepsilon$$

Consequently, $\lim_{n \to \infty} \lambda_n = \lambda$. This implies one half of the lemma.

Conversely, suppose that $\lim_{n \to \infty} \lambda_n = \lambda$ and that λ_n is an

eigenvalue of A_n for all n. Now $c_n(\tau) = \sum_{j=0}^{k} a_j^{(n)} \tau^j$ and

$$c(\tau) = \sum_{j=0}^{k} a_j \tau^j, \text{ so}$$

$$|c_n(\tau) - c(\tau)| \leq \sum_{j=0}^{k} |a_j^{(n)} - a_j| \ |\tau|^j \quad \text{for all } \tau$$

There is some $\rho \geq 1$ such that $|\lambda| \leq \rho$ and $|\lambda_n| \leq \rho$ for all
n because $\lim_{n\to\infty} \lambda_n = \lambda$. Therefore

$$|c_n(\lambda_n) - c(\lambda_n)| \leq \rho^k \left(\sum_{j=0}^{k} |a_j^{(n)} - a_j| \right) \quad \text{for all } n$$

but (a) implies that $\lim_{n\to\infty} a_j^{(n)} = a_j$ for $1 \leq j \leq k$; furthermore
$c_n(\lambda_n) = 0$; consequently, $\lim_{n\to\infty} c(\lambda_n) = 0$. On the other hand,
$\lim_{n\to\infty} c(\lambda_n) = c(\lim_{n\to\infty} \lambda_n)$ because c is continuous. Therefore
$c(\lambda) = 0$. This completes the proof of the lemma. []

Proof of Gerŝgorin's Theorem: To begin with we may
assume that the radius δ_{i_j} of each $D_{i_j}(A)$ is positive if
$1 \leq j \leq s_0$. Now define $A(\tau)$ by putting $a_{ij}(\tau) = \tau a_{ij}$ if
$i \neq j$ and $a_{ii}(\tau) = a_{ii}$ for all $0 \leq \tau \leq 1$ and all $1 \leq i, j \leq k$.
Let $\Lambda(\tau)$ denote the set of eigenvalues of $A(\tau)$. Note that
$\Lambda(0) = \{a_{i_j,i_j} : 1 \leq j \leq s\}$.

1. For some $0 < \alpha \leq 1$ we have $\Lambda(\tau) \cap S \neq \emptyset$ for all $0 < \tau \leq \alpha$.

Otherwise, for every n there would exist a τ_n such that

2. a. $\Lambda(\tau_n) \cap S = \emptyset$

    b. $0 < \tau_n \leq \frac{1}{n}$

But then the lemma would imply that $|a_{i_1 i_1} - \lambda_n| < \delta_{i_1}$ for

some n [because (2b) implies that $\lim_{n \to \infty} A(\tau_n) = A(0)$ and
$a_{i_1 i_1}$ is an eigenvalue of $A(0)$] which would in turn imply
that $\lambda_n \in \Lambda(\tau_n) \cap D_{i_1 i_1}(A) \subseteq \Lambda(\tau_n) \cap S$, contradicting (2a).

If we let $G = \{0 < \alpha \leq 1 : \Lambda(\tau) \cap S \neq \emptyset$ for all
$0 < \tau \leq \alpha\}$ and $\mu = \text{lub}(G)$, then

3. $\Lambda(\tau) \cap S \neq \emptyset$ for all $0 < \tau < \mu$

Otherwise, $\Lambda(\beta) \cap S = \emptyset$ for some $\beta < \mu$, which would
imply that β is an upper bound of G [for if $\gamma \geq \beta$ and $\gamma \in G$,
then $\Lambda(\tau) \cap S \neq \emptyset$ for all $\tau \leq \gamma$ including $\tau = \beta$]. Therefore
$\Lambda(\mu - (1/n)) \cap S \neq \emptyset$ for all sufficiently large n. Since S
is compact [see, e.g., Knopp (1952, Bolzano-Weierstrass
theorem, p. 73)] there exist an infinite sequence of
eigenvalues λ_{n_m} of $A(\mu - (1/n_m))$ converging to a point λ of
S. The lemma then implies that λ must be an eigenvalue of
$A(\mu)$, and hence

4. $\Lambda(\mu) \cap S \neq \emptyset$

If $\mu < 1$, then when $\mu + \frac{1}{n} < 1$ there would be some σ_n in
$(\mu, \mu + \frac{1}{n}]$ for which $\Lambda(\sigma_n) \cap S = \emptyset$ [otherwise $\Lambda(\tau) \cap S \neq \emptyset$
for every point τ of $(\mu, \mu + \frac{1}{n}]$ which would put $\mu + (1/n)$
in G contrary to the definition of μ]. We would then have
$\Lambda(\sigma_n) \subseteq T$ for an infinite sequence of points $\{\sigma_n\}$ converging
to μ. Since T is compact, an argument similar to that of the
previous paragraph would show that $\Lambda(\mu) \cap T \neq \emptyset$, and hence
[by (4)] $S \cap T$ would not be empty contrary to our hypothesis.
Therefore $\mu = 1$, and hence [by (4)] S contains an eigenvalue
of A.   []

Corollary 1:   Any matrix whose Gersgorin discs are pair —
wise disjoint is diagonable.

Corollary 2:   Any real matrix whose Gersgorin discs are
pairwise disjoint has only real eigenvalues.

Proof: The Gersgorin discs of a real matrix are centred
on the real axis and the eigenvalues of a real matrix, being
roots of a real polynomial, occur in conjugate pairs.

Exercise

28.   Gersgorin actually presented a stronger theorem than
      the one given above: he showed that S contains as many
      eigenvalues as it contains discs (counting each eigenvalue
      as often as its algebraic multiplicity).   Prove this
      stronger version of Gersgorin's theorem.

7.   AN APPLICATION OF GERSGORIN'S THEOREM: KY FAN'S THEOREM

Ky Fan's Theorem (1958):   If A is any $k \times k$ complex
matrix, $b_{ij} \geq |a_{ij}|$ for all i, j and λ is any eigenvalue of
A, then there is some i such that

$$|\lambda - a_{ii}| \leq |B| - b_{ii}$$

Proof: B is a nonnegative matrix.   Suppose $B > 0$.   We
have $Bv = |B|v$ for some $v > 0$ by Perron's theorem.   Let
$V = \overset{k}{\underset{i=1}{\oplus}} v_i$ and $C = V^{-1}AV$.   For some i we have

$$|\lambda - a_{ii}| \leq \sum_{j \neq i} |c_{ij}|$$

by Gersgorin's theorem.   But $c_{ii} = a_{ii}$ for all i, therefore

$$|\lambda - a_{ii}| \le v_i^{-1}\left(\sum_{j=1}^{k} b_{ij}v_j - b_{ii}v_i\right)$$

$$\le v_i^{-1}(|B|v_i - b_{ii}v_i)$$

$$\le |B| - b_{ii}$$

If B is not positive let $b_{ij}^{(n)} = b_{ij} + (1/n)$ for all i, j, then $B_n > 0$, so for some $i_n \le k$ we have

$$|\lambda - a_{i_n i_n}| \le |B_n| - b_{i_n i_n}^{(n)}$$

as we have just seen.

Now for some i between 1 and k we must have $i = i_n$ for infinitely many n because all i_n are between 1 and k. Therefore,

$$|\lambda - a_{ii}| \le |B_{n_m}| - b_{ii}^{(n_m)}$$

for some infinite sequence $n_1, n_2, \ldots, n_m, \ldots$ We saw in the proof of theorem 5 of Chap. 2 that $\lim_{n\to\infty} |B_n| = |B|$ and we also know that $\lim_{n\to\infty} b_{ii}^{(n)} = b_{ii}$, so $\lim_{m\to\infty} |B_{n_m}| = |B|$ and $\lim_{m\to\infty} b_{ii}^{(n_m)} = b_{ii}$, and hence $|\lambda - a_{ii}| \le |B| - b_{ii}$. []

Exercise

29. Prove that if A is a real matrix, $a_{ij} \ge 0$ for all $i \ne j$ and Av < 0 for some v > 0, then A is a stability matrix. (*Hint:* Consider $b_{ij} = v_i^{-1}a_{ij}v_j$.)

8. GENERALIZATIONS OF GERSGORIN'S THEOREM: OSTROWSKI'S THEOREM

Lemma: If σ and τ are nonnegative real numbers and $0 \le \alpha \le 1$, then

$$\tau^{\alpha}\sigma^{1-\alpha} \le \alpha\tau + (1 - \alpha)\sigma$$

Proof: Without loss of generality, we may assume that $0 \le \tau < \sigma$ and $0 < \alpha < 1$. Let $x = \tau/\sigma$ and $f(x) = x^{\alpha} - \alpha x$ for $0 \le x \le 1$. Since $f'(x) = \alpha(x^{\alpha-1} - 1) > 0$ for $x < 1$ and $f'(1) = 0$, we have $x^{\alpha} - \alpha x \le 1 - \alpha$ from which the inequality follows when we replace x by τ/σ.[]

Holder's Inequality: If y and z are nonnegative elements of $\underline{R}^k$ and $0 < \alpha < 1$, then

$$\sum_{j=1}^{k} y_j z_j \le \left(\sum_{j=1}^{k} y_j^{1/\alpha}\right)^{\alpha}\left(\sum_{j=1}^{k} z_j^{1/(1-\alpha)}\right)^{1-\alpha}$$

Proof: Without loss of generality, we may assume that y and z aren't 0. Applying the lemma we have just proved and taking $\tau = y_\ell^{1/\alpha}/\sum_j y_j^{1/\alpha}$ and $\sigma = z_\ell^{1/1-\alpha}/\sum_j z_j^{1/1-\alpha}$, we obtain:

$$\left(\frac{y_\ell^{1/\alpha}}{\sum_{j=1}^{k} y_j^{1/\alpha}}\right)^{\alpha}\left(\frac{z_\ell^{1/(1-\alpha)}}{\sum_{j=1}^{k} z_j^{1/(1-\alpha)}}\right)^{1-\alpha} \le \frac{\alpha y_\ell^{1/\alpha}}{\sum_{j=1}^{k} y_j^{1/\alpha}} + \frac{(1-\alpha)z_\ell^{1/(1-\alpha)}}{\sum_{j=1}^{k} z_j^{1/(1-\alpha)}} \qquad (2)$$

Summing (2) on ℓ, we obtain

$$\frac{\sum\limits_{\ell=1}^{k} y_\ell z_\ell}{\left(\sum\limits_{j=1}^{k} y_j^{1/\alpha}\right)^\alpha \left(\sum\limits_{j=1}^{k} z_j^{1/(1-\alpha)}\right)^{1-\alpha}} \leq \alpha 1 + (1-\alpha)1 = 1$$

and hence

$$\sum\limits_{j=1}^{k} y_j z_j \leq \left(\sum\limits_{j=1}^{k} y_j^{1/\alpha}\right)^\alpha \left(\sum\limits_{j=1}^{k} z_j^{1/(1-\alpha)}\right)^{1-\alpha} \quad []$$

If A is any k × k complex matrix let $\delta_i(A)$ be the radius of its ith Gerŝgorin disc. that is, $\delta_i(A) = \sum\limits_{j \neq i} |a_{ij}|$.

Theorem (Ostrowski 1951): Suppose A is any k × k complex matrix and $0 \leq \alpha \leq 1$. If λ is any eigenvalue of A, then for some $1 \leq i \leq k$

$$|\lambda - a_{ii}| \leq \delta_i^\alpha(A)\delta_i^{1-\alpha}(A^{tr}) \tag{3}$$

Proof: Since $\alpha = 0$ or $\alpha = 1$ is Gerŝgorin's theorem, we may assume $0 < \alpha < 1$. $(\lambda I - A)x = 0$ for some $x \neq 0$. Therefore for all $i \leq k$,

$$|\lambda - a_{ii}||x_i| = |\sum\limits_{j \neq i} a_{ij}x_j| \leq \sum\limits_{j \neq i} |a_{ij}||x_j|$$

Suppose (3) doesn't hold for all i. Letting

$$\Omega_i = \delta_i^\alpha(A)\delta_i^{1-\alpha}(A^{tr})$$

we have

$$\Omega_i < |\lambda - a_{ii}| \text{ for all } i \leq k \tag{4}$$

As $x \neq 0$ the set $J \equiv \{i \leq k : x_i \neq 0\}$ is nonempty and we have

$$\Omega_i |x_i| < \sum_{j \neq i} |a_{ij}|^\alpha |a_{ij}|^{1-\alpha} |x_j| \quad \text{all } i \in J \quad (5)$$

Applying Hölders inequality with $y_j = |a_{ij}|^\alpha$, $z_j = |a_{ij}|^{1-\alpha} |x_j|$, we obtain

$$\delta_i^\alpha(A) \delta_i^{1-\alpha} (A^{tr}) |x_i|$$

$$< \left(\sum_{j \neq i} |a_{ij}|^{\alpha/\alpha} \right)^\alpha \left(\sum_{j \neq i} |a_{ij}|^{(1-\alpha)/1-\alpha} |x_j|^{1/1-\alpha} \right)^{1-\alpha} \quad (6)$$

for all $i \in J$.   Therefore

$$\delta_1^{1-\alpha}(A^{tr}) |x_i|$$

$$< \left(\sum_{j \neq i} |a_{ij}|^{(1-\alpha)/(1-\alpha)} |x_j|^{1/(1-\alpha)} \right)^{1-\alpha} \quad \text{all } i \in J \quad (7)$$

because (6) implies $\delta_i(A) \neq 0$, and hence $\delta_i(A) > 0$.   We then have

$$\delta_i(A^{tr}) |x_i|^{1/(1-\alpha)} < \sum_{j \neq i} |a_{ij}| |x_j|^{1/(1-\alpha)} \quad \text{all } i \in J \quad (8)$$

The fact that $J \neq \phi$ and inequality (8) imply

$$\sum_{i=1}^k \delta_i(A^{tr}) |x_i|^{1/(1-\alpha)} < \sum_{i=1}^k \sum_{j \neq i} |a_{ij}| |x_j|^{1/(1-\alpha)} \quad (9)$$

But

$$\sum_{i=1}^k \sum_{j \neq i} |a_{ij}| |x_j|^{1/(1-\alpha)} = \sum_{j=1}^k \sum_{i \neq j} |a_{ij}| |x_j|^{1/(1-\alpha)}$$

$$= \sum_{j=1}^{k} \sum_{i \neq j} |a_{ij}||x_j|^{1/(1-\alpha)}$$

$$= \sum_{j=1}^{k} \delta_j(A^{tr})|x_j|^{1/(1-\alpha)}$$

$$= \sum_{i=1}^{k} \delta_i(A^{tr})|x_i|^{1/(1-\alpha)}$$

since i and j are dummy variables. This contradicts (9). Therefore (3) must hold for some i establishing Ostrowski's theorem. []

The following list of corollaries all apply to any k × k complex matrix A and any real α such that $0 \leq \alpha \leq 1$. Corollaries 1-3 are all due to Ostrowski (1951). They are improvements of the results listed as corollaries 4-7.

Corollary 1: If A is singular, then for some $1 \leq i \leq k$:

a. $|a_{ii}| \leq \delta_i^{\alpha}(A)\delta_i(A^{tr})^{1-\alpha}$

b. $|a_{ii}| \leq \alpha\delta_i(A) + (1-\alpha)\delta_i(A^{tr})$

Corollary 2:

a. $|A| \leq \max_{i}\{|a_{ii}| + \delta_i^{\alpha}(A)\delta_i^{1-\alpha}(A^{tr})\}$

b. $|A| \leq \max_{i}\{|a_{ii}| + \alpha\delta_i(A) + (1-\alpha)\delta_i(A^{tr})\}$

Corollary 3: If $\rho_i(A) = \sum_{j=1}^{k}|a_{ij}|$, then $|A| \leq$ $\max_{i} \rho_i^{\alpha}(A)\rho_i^{1-\alpha}(A^{tr})$.

Proof: Use corollary 2(a) and Hölder's inequality with

$$y = [|a_{ii}|^{\alpha}, \delta_i^{\alpha}(A)] \quad \text{and} \quad z = [|a_{ii}|^{1-\alpha}, \delta_i^{1-\alpha}(A^{tr})] \;[]$$

Corollary 4: [A. B. Farnell (1944)]

$$|A| \leq \max_i \sqrt{\rho_i(A)\rho_i(A^{tr})}$$

Corollary 5: [A. Brauer (1946)]

$$|A| \leq \min(\max_i \rho_i(A), \ \max_i \rho_i(A^{tr}))$$

(This is Exercise 10, Chap. 2 again: use corollary 3 with appropriate α to prove it.)

Corollary 6: [W. V. Parker (1937)]

$$|A| \leq \max_i \frac{1}{2}\left(\rho_i(A) + \rho_i(A^{tr})\right)$$

Corollary 7: [E. T. Browne (1930)]

$$|A| \leq \frac{1}{2} [\max_i \rho_i(A) + \max_i \rho_i(A^{tr})]$$

Exercise

30. Use Ostrowski's theorem and/or its corollaries to estimate $|A|$ if A is as in Exercise 18. In particular, corollary 3 with $\alpha = \frac{1}{2}$.

Ostrowski also was able to generalize Gersgorin's theorem in another way. Define an *oval of Cassini of* A by

$$O_{ij}(A) = \{z \ \epsilon \ \underline{C} : \ |z - a_{ii}||z - a_{jj}| \leq \delta_i(A)\delta_j(A)\}$$

for all $i \neq j$ between 1 and k.

Theorem (Ostrowski 1937): Every eigenvalue λ of an arbitrary $k \times k$ complex matrix A lies in some $0_{ij}(A)$ (assuming $k \geq 2$).

Proof: $(\lambda I - A)x = 0$ for some $x \neq 0$. Choose r and t so that $r \neq t$ and $|x_r| \geq |x_t| \geq |x_j|$ for all $j \neq r$.

If $x_t = 0$, then all entries in x except x_r are zero, and hence

$$\lambda x_r = \sum_{j=1}^{k} a_{rj}x_j = a_{rr}x_r$$

Since $x_r \neq 0$, we have $\lambda = a_{rr}$, and hence

$$|\lambda - a_{rr}||\lambda - a_{rt}| = 0 \leq \delta_r(A)\delta_t(A)$$

Suppose $x_t \neq 0$. We have

$$(\lambda - a_{ii})x_i = \sum_{j \neq i} a_{ij}x_j \quad \text{for all } i$$

therefore

$$|\lambda - a_{rr}||x_r| \leq \sum_{j \neq r} |a_{rj}||x_j| \leq |x_t|\delta_t(A)$$

and

$$|\lambda - a_{tt}||x_t| \leq \sum_{j \neq t} |a_{tj}||x_j| \leq |x_r|\delta_t(A)$$

Therefore $|\lambda - a_{rr}||\lambda - a_{tt}||x_r||x_t| \leq |x_r||x_t|\delta_r(A)\delta_t(A)$. As $x_t \neq 0$ and $x_r \neq 0$, it follows that

$$|\lambda - a_{rr}||\lambda - a_{tt}| \leq \delta_r(A)\delta_t(A)$$

Consequently, whether or not $x_t = 0$, we have found distinct r and t such that

$$\lambda \ \varepsilon \ O_{rt}(A) \ []$$

Corollary 1:   If A is any $k \times k$ complex matrix $(k > 2)$ and $|a_{ii}||a_{jj}| > \delta_i(A)\delta_j(A)$ for all $i \neq j$, then A is nonsingular.

Example: If

$$A = \begin{bmatrix} 2.0 & 1.1 & 1 \\ -0.8 & 3.0 & 2 \\ 1.2 & 1.1 & 3 \end{bmatrix}$$

then

$$(\delta_1(A), \ \delta_2(A), \ \delta_3(A)) = (2.1, \ 2.8, \ 2.3)$$

so

$$|a_{11}||a_{22}| = 6 > (2.1)(2.8)$$

$$|a_{11}||a_{33}| = 6 > (2.1)(2.3)$$

$$|a_{22}||a_{33}| = 9 > (2.8)(2.3)$$

and hence A is nonsingular. Notice that the test for nonsingularity based on Gerŝgorin's theorem (Exercise 25) does not help for this matrix.

Exercise

31. Ostrowski's theorem of 1937 might lead you to believe that:

   a. If λ is an eigenvalue of A and A is an arbitrary $k \times k$ matrix $(k \geq 3)$, then

$$|\lambda - a_{ii}||\lambda - a_{jj}||\lambda - a_{tt}| \leq \delta_i(A)\delta_j(A)\delta_t(A)$$

for some triple of distinct indices i, j, t. Show

that this is not true by considering $A = \begin{bmatrix} 1 & 0 & 0 \\ 0 & 1 & 1 \\ 0 & 1 & 1 \end{bmatrix}$

b. Show that if $3 \leq \ell \leq k$, then there is some k × k

matrix A at least one of whose eigenvalues satisfies

$$|\lambda - a_{i_1 i_1}||\lambda - a_{i_2 i_2}| \cdots |\lambda - a_{i_\ell i_\ell}| >$$

$$\delta_{i_1}(A)\delta_{i_2}(A) \cdots \delta_{i_\ell}(A)$$

for every ℓ-tuple of distinct indices $(i_1, i_2, \ldots, i_\ell)$.

This exercise shows that no further extension of
Ostrowski's theorem is possible along the lines indicated.
However, there is another type of extension which is possible.

Theorem (Ostrowski): If A is any k × k complex matrix
$(k \geq 2)$, λ is any eigenvalue of A, and $0 \leq \alpha \leq 1$, then

$$|\lambda - a_{ii}||\lambda - a_{jj}| \leq \left[\delta_i(A)\delta_j(A)\right]^\alpha \left[\delta_i(A^{tr})\delta_j(A^{tr})\right]^{1-\alpha}$$

for some distinct pair of indices (i, j).

If you are interested in the proof, it is given in
Marcus and Minc (1964, Chap. 8, Sec. 2.56). The original
paper by Ostrowski appears in Compositio Math., vol. 9 (1951),
pp. 209-226.

MISCELLANEOUS EXERCISES

1. Does $\sum\limits_{n=0}^{\infty} \frac{n}{6^n} \begin{bmatrix} 1 & -8 \\ -2 & 1 \end{bmatrix}^n$ converge? Explain.

2. Evaluate $\sum\limits_{n=0}^{\infty} \dfrac{1}{3^n} \begin{bmatrix} 1 & 2 \\ 1 & 1 \end{bmatrix}^n$.

3. Evaluate $\sin \begin{bmatrix} \pi & 1 & 0 & 0 & 0 \\ 0 & \pi & 1 & 0 & 0 \\ 0 & 0 & \pi & 0 & 0 \\ 0 & 0 & 0 & 0 & 1 \\ 0 & 0 & 0 & 0 & 0 \end{bmatrix}$

4. If $M^2 = M$, show that M is diagonable and describe its eigenvalues.

5. If $\lim\limits_{n\to\infty} M^n = M$, show that trace(M) = rank(M).

6. What can you say about the Jordan form of a

    a. power-bounded matrix?

    b. power-convergent matrix?

7. Find $\arctan \begin{bmatrix} 1 & 0 & 0 \\ 0 & 0 & 1 \\ 0 & 0 & 0 \end{bmatrix}$.

8. If A is a k × k complex matrix and $\lim\limits_{n\to\infty} A^n$ is nonsingular, show that A = I.

9. Suppose P is a stochastic matrix.

    a. Show that P is power-bounded.

    b. Prove that the sequence P, $((1/2)(P + P^2))$, ... $(1/n) \sum\limits_{j=1}^{n} P^j$, ... has a finite limit. [*Hint:* Use the similarity method and (a).]

    c. If P is the transition matrix of a Markov chain whose initial vector is $p^{(0)}$, how would you interpret the ith entry in $p^{(0)}[\lim\limits_{n\to\infty} (1/n) \sum\limits_{j=1}^{n} P^j]$?

10. Suppose the characteristic polynomial of 5 × 5 complex matrix A is $(\tau - 1)^2(\tau - \frac{1}{2})^2(\tau - \frac{1}{4})$. Show that A is power-convergent if, and only if, the rank of A - I is 3.

11. If A is any nonnegative matrix prove that:

   a. Its spectral radius is an eigenvalue having a
       nonnegative eigenvector. [Suggestion: Use the
       Frobenius normal form of A (Exercise 35, Chap. 2).]

   b. If λ is an eigenvalue corresponding to a positive
       eigenvector, then λ the spectral radius of A.
       [*Hint:* Look at the proof of the corollary to
       theorem 1, Chap. 2.]

12. Supply an example of the matrix or matrices described
    below if such an example exists. If there is no such
    example explain briefly why not.

   a. Two 2 × 2 matrices which are not similar but have
       the same characteristic polynomial.

   b. A 2 × 2 matrix whose spectral radius is not one of
       its eigenvalues.

   c. A positive matrix similar to $\begin{bmatrix} 1 & 0 \\ 0 & 0 \end{bmatrix}$.

   d. A positive matrix similar to $\begin{bmatrix} 2 & 2 & 1 \\ 0 & 2 & 1 \\ 0 & 0 & 1 \end{bmatrix}$.

   e. A nonnegative matrix similar to $\begin{bmatrix} -2 & 2 & 3 \\ 0 & 1 & 1 \\ 0 & 0 & 1 \end{bmatrix}$.

13. Evaluate $\lim_{n \to \infty} (A/|A|)^n$ if $A = \begin{bmatrix} 1 & 2 & 1 \\ 2 & 4 & 2 \\ 1 & 2 & 1 \end{bmatrix}$.

14. Suppose A is a k × k real matrix and $f(\tau) = \sum_{n=0}^{\infty} \alpha_n \tau^n$
    converges for all scalars τ. If all $\alpha_n \geq 0$, show that
    $|f(A)| \leq f(|A|)$ with equality if $A > 0$.

15. Suppose A is a k × k real matrix and $a_{ij} > 0$ for all
    $i \neq j$. If α is the largest of the real parts of the
    eigenvalues of A, show that α is a simple eigenvalue

of A having a positive eigenvector. (Suggestion: Choose
a scalar σ so that $\sigma I + A > 0$, then compare the spectrum
of $\sigma I + A$ to the spectrum of A.)

16. If A is an n × n matrix and sin 2A is singular, show
that sin A or cos A is also singular.

17. If $dX/dt = \begin{bmatrix} \cos t & 1 \\ t^2 & \cos t \end{bmatrix} X$ for all t and $X(0) = \begin{bmatrix} 2 & 2 \\ 2 & 3 \end{bmatrix}$,
find det $X(\pi)$.

18. If A is nonnegative and its row and column sums all
equal the same positive number α, prove that $|A|^{-1}A$ is
power-bounded.

19. If $A \geq 0$, to what extent is it true that there exists
a diagonal matrix D and a stochastic matrix S such that
$A = D^{-1}SD$?

20. Without any lengthy computations

a. Show that $\begin{bmatrix} -3 & 1 & -1 & 0 & 0 \\ 2 & -5 & 1 & 0 & 1 \\ -1 & 0 & -4 & 1 & 1 \\ 0 & 1 & 0 & -2 & 0 \\ 0 & 1 & -1 & 1 & -4 \end{bmatrix}$ is a stability matrix.

b. Show that $\sum\limits_{n=1}^{\infty} (1/n)A^n$ diverges if $A = \begin{bmatrix} 4 & 1 & -1 & 0 \\ 1 & 4 & 0 & -1 \\ 0 & -2 & 4 & 0 \\ 2 & -1 & 1 & 4 \end{bmatrix}$

21. Suppose A and B are complex k × k matrices and AB = BA.

a. Show that A and B have a common eigenvector.
     (*Hint:* Let λ be an eigenvalue of B and W =
     $\{x \in \underline{C}^k : Bx = \lambda x\}$, show that A maps the subspace
     W into W and consider the restriction of A to W.)

b. Show that there exist upper triangular S and T and a unitary U such that both $A = U^*TU$ and $B = U^*SU$. (*Hint:* Use (a), even if you haven't proved it, and mimic the proof of Schur's theorem.)

c. If A and B are also Hermitian, show that there exist diagonal matrices D and E and a unitary matrix U such that both $A = U^*DU$ and $B = U^*EU$. (*Hint:* Use (b), even if you haven't proved it.)

22. If A is a real symmetric k × k matrix, show that
$$\frac{1}{k} \sum_{i=1}^{k} \sum_{j=1}^{k} a_{ij} \leq |A|.$$

23. Suppose A is a k × k real matrix. If $a_{ii} < -\sum_{j \neq i} |a_{ij}|$ for all $1 \leq i \leq k$, show that A is a stability matrix.

24. If $A = \begin{bmatrix} 6 & 5 & 1 & 2 \\ 1 & 7 & 0 & 2 \\ 0 & 4 & 7 & 5 \\ 2 & 0 & 1 & 5 \end{bmatrix}$, show that $|A| < 13$.

REFERENCES

BELLMAN, R.:   An Introduction to Matrix Analysis, 2nd ed.,
                McGraw-Hill Book Co., New York, 1970.

CHUNG, K.L.:   Markov Chains with Stationary Transition
                Probabilities, Springer-Verlag, Berlin, 1960.

CHURCHILL, R.V.:   Introduction to Complex Variables and
                    Applications, McGraw-Hill Book Co., New York
                    1948.

FELLER, W.:   An Introduction to Probability Theory and its
               Applications, vol. I, 2nd ed., John Wiley and
               Sons, New York, 1957.

FRANKLIN, J.N.:   Matrix Theory, Prentice Hall, New Jersey,
                   1968.

FULKS, W.:   Advanced Calculus, John Wiley & Sons, New York,
              1961.

GANTMACHER, F.R.:   Applications of the Theory of Matrices,
                     Vols. I and II; (translated by K.A. Hirsch),
                     Chelsea, New York, 1959.

HAHN, W.:   Stability of Motion, Bd 138 Series: Die Grundlehren
             der mathematischen Wissenschaften, Springer-Verlag,
             Berlin, 1967.

HOFFMAN, K. AND KUNZE, R.:   Linear Algebra, Prentice Hall, New
                              Jersey, 1961.

HOLLADAY, J.C. AND VARGA, R.S.:   Proc. Amer. Math. Soc., Vol.
                                   9, 1958.

KEMENY, J.G. AND SNELL, J.L.:   Mathematical Models in the Social
                                 Sciences, Blaisdell Publ. Co., New York, 1962.

KNOPP, C.:   Elements of the Theory of Functions, Dover Publ.
             Co., New York, 1952.

LANCASTER, P.:   Theory of Matrices, Academic Press, New
                 York, 1969.

LIPSCHUTZ, S.:   Theory and Problems of Linear Algebra,
                 Schaum's Outline Series, McGraw-Hill Book Co.,
                 New York, 1968.

MARCUS, M. AND MINC, H.:   A survey of Matrix Theory and Matrix
                 Inequalities, Allyn and Bacon, Inc., Boston, 1964.

ROSENBLATT, D.:   U.S. Naval Logistics Quarterly, Vol. 9, 1958.

SCHUTZ, B.E.:   Motion of a Spacecraft Near a Triangular
                Libration Point of the Earth-Moon System,
                Engineering Mechanics Research Laboratory,
                University of Texas, Austin, 1966.

SENETA, E.:   Nonnegative Matrices, George Allen and Unwin
              Ltd., London, 1973.

WHITTAKER, E.T.:   A Treatise on the Analytical Dynamics of
                 Particles and Rigid Bodies, 4th ed., 1936,
                 Dover Publ. Co. reprint, New York, 1944.

WIELANDT, H.:   Topics in the Analytic Theory of Matrices,
                Lecture Notes prepared by R.R. Meyer, University
                of Wisconsin, Madison, 1967.

Suggested Collateral Reading

Any of Bellman, Gantmacher, Franklin, Lancaster, and
Marcus and Minc will serve as a useful adjunct to most of
the material in Chapters 1 through 4.   In addition, Hoffman
and Kunze, Lipschutz, and Wielandt are useful for Chapter 1.
Chung, Feller, Kemeny and Snell, Seneta, and Wielandt are
helpful for Chapter 2.

NOTATION

In this index: A and B are square matrices; x and y are vectors; n, m are integers; M is a matrix or a vector, and τ is a scalar.

SYMBOL	DESCRIPTION	PAGE OF FIRST OCCURANCE
$\underline{C}$	the complex numbers	199
$\underline{R}$	the real numbers	126
$\underline{Z}$	the integers	112
$\underline{C}^n$	the complex n-tuples	58
$\underline{R}^n$	the real n-tuples	85
Ω	a (possibly infinite) interval on the real line containing zero	127
$\oplus$	direct sum	11
$\binom{n}{m}$	the n, mth binomial coefficient	17
I_n	the n × n identity matrix	37
$J_n(\tau)$	the n × n Jordan block with eigenvalue τ	11
0_n	the n × n zero matrix	38
U_n	the n × n Jordan block with eigenvalue 0	42
x × y	the vector cross product of x by y	193
x · y	the dot product of x and y	200

INDEX